SUPERCHARGE
YOUR BUSINESS
WITH TECHNOLOGY

SUPERCHARGE YOUR BUSINESS WITH TECHNOLOGY

THE SMART WAY BUSINESS OWNERS LEVERAGE TECHNOLOGY TO DRIVE BUSINESS RESULTS

DAVID QUICK

total cover IT.

Published by Total Cover IT®
https://www.totalcoverit.com

Graphics designed by Heather Felty/Inner Peace Press

*To my mother, who ingrained in me
to always be the best at what I do,
and my father, who inspired me with
his quiet courage.*

Table of Contents

Introduction

Never did it occur to me that I would write an entire book. I never liked writing. I always dreaded English class from elementary through high school. Ultimately, I had to go to a remedial English class when I started college. Fortunately, I had a great teacher there and quickly picked up on it and progressed quickly to regular classes and received excellent grades in writing. Still, I thought writing to be an arduous task, which is why I focused on numbers-based studies in college such as mathematics and ultimately accounting.

After college, writing took a backseat as I began my professional life. My first full-time job was as an accountant for a local community bank. I liked it because I liked working with numbers and that seemed to be the ideal scenario for me. Computers were something of a hobby of mine at the time. It was the early days of the computer revolution. I remember how great I thought my Commodore VIC-20 was, getting games to run on it. Then I got my first PC computer, a 486-based computer that I thought was the greatest thing since sliced bread. There was always something new and interesting happening in technology.

The idea of applying technology solutions to business problems was with me even back then. At my accounting job, I worked manual systems based on pencil and paper! Over time, I moved all these systems to computer spreadsheets, greatly making my job much easier. However, a byproduct of that was I became TOO efficient. I quickly grew bored with my job and accounting in general. There was nowhere to grow in my role. Circumstances eventually took their course, and the bank was sold. By that time, I wanted to information technology (IT) and found my transition job — a job as accountant and system administrator at a local printing company. That was the stepping off point in my IT career. I eventually transitioned completely to IT/technology in future roles.

It seemed so hard making the transition from accounting to IT at the time. From my vantage point as of this writing in 2024, looking back over 25+ years of experience in IT, I have seen all kinds of setups, both in job experience and consulting work, generally in small business situations. Much of it was shocking. In one scenario, I was looking at a company's network equipment and seeing cables flying all over the place like spaghetti, wondering how all that continues to work. They never engaged with us, deciding the cheap way was the better way. Sure enough, I learned afterward that their network had a major crash (no surprise!), and they called their current IT consultant (who had a day job in another state!). They were willing to wait for him to get time off and make the 4-5 hour drive to their location while their systems were down! It was mind-blowing!

Small businesses struggle. Day-to-day getting-the-work done reigns supreme and concerns about supporting technology go by the wayside. I wanted to help small businesses to do better and educate them on the benefits of looking at IT in a different way, a strategic way, one that will, in fact, yield financial benefits to their business. But, how to get the message across? I gave this thought for quite a long time, while observing various disaster network setups over the years. It was disappointing to see that so many small businesses typically did not get things right with technology, not realizing that they can do something that would in fact help them to run their businesses better and go farther.

I do not know where the inspiration for this book finally came from. Although I was relatively skilled at writing now, writing a whole book seemed to be only a remote possibility. Perhaps it was seeing my business coach, a person who I personally knew, write his book. In any case, it took me close to three years to finally complete writing this book, overcoming the distractions of the day-to-day challenges of running a business, and perhaps my latent distaste for writing in general. It has been quite a long journey. I wrote this for you, the small business owner. I hope you find it useful.

Chapter 1
The Cost-Based View of Information Technology in a Business

When you look at small businesses, they are typically very cost conscious, and rightly so. Why? Because they tend to have more limited resources than larger businesses and they need to make wise use of their resources. Small businesses that need computers and related technology tend to work to get equipment and services at minimal cost. It's very simple. Their philosophy often goes something like this: the less I spend, the more I keep. That is the long and the short of the cost-based view. This is not necessarily just with IT. Business owners who think this way typically will carry this forward to most other products and services that they buy. For many products and services, it is fine to use the cost-based approach, particularly with products that are considered a "commodity." Paper towels are paper towels — there may be some differences between brands, but they more or less do the same thing, so you generally will shop it out for the lowest cost irrespective of the brand. However, it is not always the best approach, particularly with high value products and services that have a big impact on the business. In those cases, cheaper is not always better.

It is particularly concerning that IT is commonly viewed from a cost perspective, especially in light of the growing importance of IT in operating a modern business. For example, you may not necessarily want to buy a printer meant for a low-volume home office environment if you are looking to print thousands of pages of documents on a regular basis — you would likely want a commercial grade laser printer designed for that purpose. Spending several thousand dollars to have relatively trouble free and efficient printing versus a few hundred dollars on a printer that will likely break down or "hiccup" frequently with relatively slow output (which would be highly disruptive to your business workflow) just makes good business sense.

When you think about it, practically every business uses computers in some way. Even when you walk into a small retail shop, they typically will have at least one computing device for point-of-sale purchases. Today's cash registers are in fact computers. Other businesses may have more extensive computing technology in place. For example, with your typical Certified Public Accountant (CPA) firm today, every employee will generally have an assigned computer that includes all of the productivity and accounting applications.

This was not always the case. In the early days of computing, systems were extraordinarily large, super expensive, and were primarily used in the realm of governments, universities, and large corporations. Computers started to make their way into mainstream small businesses in the 1980s and 1990s. I remember the first PC that I worked on in an office, in 1989. It was an IBM PC, with a 286 processor as I recall, and had cost the company I worked for about $10,000! By today's standards, it had a fraction of the power of our modern computers. However, it was innovative at the time. It was ultimately replaced within a few years by a "clone" at a much lower price point. The new computer boasted a 386 processor; it was much more powerful, and, boy, I thought it was great to have a much newer and more powerful computer! And I was able to run this new thing called Windows® and have a graphical interface on a color monitor, instead of that difficult-to-use DOS (Disk Operating System) on an ugly monochrome green monitor!

It is illustrative of the progression of the market in general. IBM blazed the trail in the PC market, but ultimately found themselves outmaneuvered by rivals with "IBM Compatible" clones at a much lower cost, which made it more affordable for small businesses. Now, computer technology is commonplace everywhere. It is amazing to think that a typical smartphone today is far more powerful than that clunky IBM I worked on in my first job. Even our flash drives today can store more data than those computers! It just goes to show how far computing technology has progressed and how fundamental it is to everything we do today.

When we look at buying computer equipment for a small business, business owners tend to lose sight of the significance of how it impacts their businesses. Typically, he or she is going to evaluate them primarily on costs, not on what it can do for their business. The same goes for services. Why hire consultants when I could get my kid or my employee's kid to do it for nothing? Or maybe I'll just do it myself! The high-value business owner taking time away from his or her work to troubleshoot a computer problem? This will diminish the overall profitability of the business in nearly every case. That is the cost-based view of IT. And what you need to consider here is that in the cost-based view of IT, value from that service is not considered, nor is the impact of buying less expensive equipment or services, or simply doing without them. There is no consideration as to what that technology will do for the business. It is simply evaluated on costs. Cost is the driving factor. It equates information technology to, say, the person who waters your flowers. That's an expense. Or the person who does the cleaning of your office. That's an expense. IT is viewed in the same realm as those types of services. It is a cost. It's a cost that needs to be minimized. There is no thought as to what impact it will have on the business.

"One common result of not making the proper investment in your IT is downtime."

What is it costing businesses to not hire a professional IT service? One common result of not making the proper investment in IT is downtime. Downtime may be due to a number of factors, such as equipment that did not receive proper care and maintenance, or perhaps was being used well beyond its normal lifecycle. An example would be an old server used for file sharing that normally should be replaced after five years was still being used 10 years after its installation. Now the hard drives are failing, and there was no IT company retained to monitor it. Perhaps they had somebody's kid or an intern come in once in a while as they did not want to spend the money on a managed service provider. Ooops! Now the employees have no means to access the data on the server. Someone had an idea — let's restore from the backups! Little did they know that the backups stopped working almost a year ago. No one was hired to monitor and test the backups! What impact do you think that will have on the business? And how much did they save by not hiring IT professionals and not upgrading their server? How much did they lose? Maybe they had an important deadline they needed to meet for a client which now cannot be met. Perhaps they were working on a large proposal, and now all that information is lost. Perhaps there was information on client projects that were in the files, and now all that work is lost. If they are a manufacturer, perhaps this affects their ability to deliver the finished product on time.

All of these scenarios result in angry clients/customers, loss of reputation, not to mention frustrated employees throwing up their hands — and you are paying them to sit around and do nothing!

"A business not making the investment in a document management/workflow system may be left behind."

And how about workflow? Imagine a firm circulating important paper forms around the office. Maybe it's a CPA office moving paper tax or audit files around the office for input and review. Suppose a paper gets lost? Or no one can find the file, and it is sitting on someone's desk, and they totally forgot about it? And now with the growing remote workforce, how do you move paper around? A business not making the investment in a document management/workflow system may be left behind the competition, as well as having vendors waiting for payment and clients waiting for service. If you served your clients faster, you would get paid faster, and they would be happier with your service! Perhaps you could even charge more!

And let's not forget the cybersecurity implications. There is the potential for legal exposure due to a breach, as well as compliance penalties if the business is in a regulated industry such as in healthcare or financial services. I once had a client who cringed at the mere thought of buying a firewall to protect their computer network instead of having a router which does very little for security. This was a professional services firm with valuable client information to protect, but no thought was given to spend money on even basic security.

Threat actors love businesses that don't like to spend money, because they are the low hanging fruit! If given a choice between hacking a business where there were proper investments in cybersecurity and one where there was not, they will pick the one that was not every time, to get the most bang with the least effort. They want the quick ransomware payday! We will discuss cybersecurity in greater detail in later chapters, citing case studies and how to best protect your business. The decision to "cheap out" on IT is not one to be taken lightly. The negative cascading effects to a business are endless and may potentially destroy the business.

> *How are you handling decisions on IT products and services?*

Chapter 2
Big Growth by
Bits and Bytes

When you think of the more traditional view of IT as a cost, a business is just spending as little as necessary to get the equipment/software/applications they need to run their business, as opposed to the more strategic view of IT. These are represented by two different approaches. One is simply saying, I'm just going to keep what I have in place as long as I can. Just get by with it for as long as possible, and only spend when it is absolutely, positively necessary, and spend as little as possible. Then there are those that take the opposite view — what are my business objectives and what investment do I need to make in my technology to get me there? Let's explore what happens when a business takes this "business investment" approach to technology.

A business that makes the proper investment in technology will get ahead and grow by leaps and bounds. Look around you. There are many examples of businesses that have taken that view, even if they may not necessarily explicitly communicate the fact that they've done that, you can see the results of what they do. They dominate their industry, both on the e-commerce side and Cloud business. They have made tremendous investments in their technology. And they are just leapfrogging all their competitors. They've driven other e-commerce companies out of business and severely diminished brick and mortar retailers, becoming the dominant force in their industry. The most obvious, and perhaps the best, example is Amazon®.

Just my own experience with them demonstrates how they have made these investments in technology. When I order from Amazon, it is so simple and easy. I just find what I need there. It only takes a few clicks. Then it's done; it's very simple. They've made the process so user-friendly and trouble free. I used to have some trepidation, a lot of trepidation, in fact, about ordering online, not only from Amazon, but from any e-commerce company. I was always concerned if what I ordered was not adequate, or if it was damaged, or it didn't meet my needs, that I could not return it. In the old days, I would just go to the local department store, I'd buy whatever I needed there, and be assured that if I had a problem I could bring it back — like a piece of clothing, for example. Not a problem; exchange it and be done with it. The process was very simple and straightforward. But it didn't seem like that was the case with e-commerce because you're not going into a store; you're not trying clothing on to make sure it fits you. You don't have that physical interaction with the thing that you want to buy to know for sure you're buying the right thing and at least have some safeguards that it is returnable. Now it seems like that experience has been brought into e-commerce, as well.

I will relate my buying experience with Amazon. I bought three items, two were pieces of clothing, neither of which were adequate for my needs. I also had bought a piece of computer equipment that turned out to be not what I was looking for. So, I went through the return process online, which was very simple — a few clicks of the mouse and I was given the option of dropping it off at the local Whole Foods®, which I chose. I was going there anyway to do some shopping. Right inside the entrance to Whole

Foods, just to the left, is the place where you do the return. I walked up to the counter, the attendant took each item in turn, scanned a QR code on my phone, then scanned the label she applied to the package to be sent back. She did that for each one of the three items and 1-2-3 it was done. I just walked away. To my surprise, within just a few minutes, I received confirmation that the return was done. And within another hour or two, I got confirmation that the return would be credited to my credit card and applied in a few days. Very simple. The experience was almost as good as just returning the items to a department store. It was virtually the same experience. It was interesting to me that Amazon made it almost like the same experience. It made me more inclined to buy more from Amazon because now I know that I have some easy options for returns. If I have a problem with these items I could simply physically take the item back somewhere and get credit for it. How simple is that? And multiply this by the number of people shopping at Amazon who had the same experience. That just goes to show you that the results speak for themselves. But it's just one example of how technology helps a business to naturally be more efficient, helps them to grow, helps them to overtake their competition.

It's All About the System

Probably the best example of how investing in technology can transform a business is with a software system that helps businesses run their entire business: enterprise resource planning (ERP). ERP systems are an all-encompassing application that touches every aspect of a company's operations. It encompasses things from the financial side, which would be something you commonly think of with ERP, but it's not just accounting software. It's not like QuickBooks®. QuickBooks is accounting software. ERP systems go beyond accounting software; they reach into every aspect of a company's operations and all aspects of the sales process. For example, with a manufacturer, an ERP system would integrate with the manufacturing process of that company. With a warehousing company, they would be integrating with the warehousing operations. If salespeople have insight into data that would be generated from the ERP system, that will allow them to make better planning decisions on sales.

"One of the primary values of an ERP system is that it helps an organization to break down the silos and facilitate cross-functional interaction."

Likewise, for the operational side, management will have insight into the company's operations, they will make better decisions, and will be in a better position to steer the company forward. One of the primary values of an ERP system is that it helps an organization to break down the silos and facilitate cross-functional interaction. There are a number of ERP applications out there. I'm not going to go into the validity of each and every one. There are major players, such as Microsoft® Dynamics for the SMB (server message block) market, but there are also much larger applications out there like SAP® and Oracle®, that cost exorbitant sums of money. Typically, those are in the realm of enterprise businesses. But you don't have to be an enterprise business to use an ERP system. There are plenty of ERP applications that are tailor made for smaller businesses, whether they are established businesses, growing businesses, or startup businesses.

To bring further light to this topic, and for you to better understand how an ERP system may work for you, I spoke with a couple of people who deal in this area. Let's take a look from the perspective of the ERP implementers.

I spoke with Don Waters, President of BC Systems, discussing the ways in which his ERP implementations have brought substantial business results to his clients. BC Systems had its origins through a predecessor business established in 1984, and has evolved to specialize in the implementation of cloud-based ERP systems. The ERP system that they generally implement is a cloud-based solution named Priority Cloud ERP. One example is an implementation they did for a testing company client. The client's business involved testing metals to ensure accurate composition. Prior to the implementation of Priority, it was a manual process. Because it was manual, errors were introduced, and manual corrections had to be made to the results. This resulted in frequent retests that cost the company extra money which they could not bill back to their customers. They implemented Priority, which took about two years. Priority was fully integrated with the equipment. As a result, the test results were more accurate, and they were verifiable certifiable results. No manual intervention was needed. Retests were virtually eliminated, making the workflow more profitable and this allowed them to increase the number of tests that they performed over a period of time. Overall, the ERP implementation contributed to increasing sales by 15-20% per year (excepting 2020, which was affected by the COVID-19 pandemic).

Marvin Fischer, President of GoldFinch Cloud Solutions, has spent about 30 years as of this writing in the ERP industry and has seen the evolution from on-premises ERP software to cloud-based solutions. Marvin's company, originally named Micro DataNet, was focused on the implementation of on-premises ERP solutions. Over time, major ERP software developers moved away from on-premises software in favor of cloud-based software with subscription models. This adversely affected Marvin's business, as his business model up to that time was primarily composed of selling ERP on-premises software along with the implementation of that software. Facing a potentially catastrophic challenge, Marvin and his partner Scott Zhuo decided to move the company in a new direction, with Scott developing an in-house cloud-based ERP solution. GoldFinch ERP is the end result of that development, completing the evolution in Marvin's company from on-premises ERP software to a cloud-based ERP solution. Goldfinch leverages the SalesForce® platform to provide a feature-rich and adaptable solution. Marvin discussed a few examples of how GoldFinch has brought tangible business results using this platform.

One such example was Beyond Better Foods, makers of Enlightened® ice cream and other sweet snacks. Prior to the GoldFinch implementation, every department was siloed, meaning that employees in one department did not have definitive knowledge of what was going on in other departments and what impact there was on the business. The process was largely manual, which meant that the data management process was very time consuming and was highly error prone. It was apparent that in order to grow and take them to the next level, they needed to implement an ERP system.

Implementing the new system produced significant results. It empowered salespeople with actionable data to drive more sales. Salespeople can confidently approach customers with accurate data on what promotions worked for other customers, saying something like, "So and so down the street did this promotion with us and the results were X." This was game-changing, powerful data salespeople had in their arsenal.

With the data on one integrated platform, decision-making was more efficient and effective, and the data was free of errors and was available and actionable by all departments. It is estimated that the savings from errors was roughly in the range of six figures, as well as improved results from being able to drive more sales.

Another client of Marvin's that benefited greatly from implementing GoldFinch was an HVAC vendor located in the Midwest. Their workflow included field technicians who performed work onsite. Before GoldFinch was implemented, service technicians closing out a work ticket would not record supplies used on the job, such as how many gloves, how much oil, etc. As a result, bills were not accurate, and inventory management was very disjointed — should we stock an item or not? Marvin's group learned of these issues in developing the project implementation plan. Once the implementation was complete, before they closed the job ticket, field service technicians were asked what supplies were used and had to record them in order to close out the ticket. Bills were now more accurate and inventory management was very efficient. The company estimated this change alone resulted in about $2,000,000 in additional sales.

Product delivery processes are another area where an ERP system provides benefits. Marvin had a client who was facing costly challenges when dealing with their large retail chain customers. A nuance of the retailing industry is the concept of chargebacks. A manufacturer or wholesaler sells a product to a large retailer. Delivery is made, but the retailer claims non-delivery or delivery outside of their specific time window, so the large retailer charges the manufacturer or wholesaler a penalty. These add up and are potentially very costly. This is unfortunately a common practice of large retailers. The client implemented GoldFinch, which resulted in more precise tracking of what was shipped and when. When a large retailer made a claim of non-compliance, the client produced documented proof that the shipment was made and made on time, forcing the large retailer to yield. Eventually, the large retailers just gave up and stopped trying to charge back the client. GoldFinch ultimately saved the client hundreds of thousands of dollars by eliminating the large retailer chargebacks.

The Time-Money Value of Intelligent Systems

Another technology that helps transform businesses is Customer Relationship Management (CRM) systems. Pete Romano, CEO and founder of Segwik, a CRM that includes complete management of the customer process, talks about a couple of his clients who greatly benefited by moving to the application. One such client was Majestic Awning. Before the implementation, they used a variety of different software applications to run the business. QuickBooks, spreadsheets, email, website. Leads would call them on the phone and information would be written down manually, name, address, how to get to their house, etc. All leads were tracked on index cards. It was a cumbersome, inefficient process. Segwik provided them with an application that handles the whole process, from leads, to quotes, to invoicing, to calendars, and more. For example, appointment reminders would go out automatically, both to the customer and to the salesperson. As a result of implementing Segwik, annual sales increased from $3 million to $6 million.

The CPA profession has seen a remarkable transformation in their workflows as a result of technology improvements. Let's take something as basic as doing an audit. Auditors used to go to clients' sites and they would input the information they needed into their software tools, such as spreadsheets in conjunction with special auditing software. The data would all be stored locally on the laptops themselves. If they worked in teams, they may very well set up a peer-to-peer network with their colleagues to work in collaboration and share the data with the "host" laptop. When they all got back to the office, they would need to sync back the data so it is properly stored on the internal network, so other staff and partners in the office may work on the audit data. This was a very cumbersome and inefficient process and lent itself to potential issues onsite with making the peer-to-peer setup work properly, as well as potential security issues with data sitting on user laptops. A great deal of time was wasted on technical issues and time is money with a CPA firm that has billable partners and staff unable to bill because of technical problems. You most certainly do not want the audit partner spending time troubleshooting a technical problem while at a client!

The great revolution occurred with remote access software. Perhaps one of the most common remote access applications is Citrix® (originally MetaFrame, then rebranded to Presentation Server, XenApp, then Virtual Apps and Desktop, and who knows what it's called now?), which was a preferred choice among midsize and large firms. Audit staff no longer had to deal with the peer-to-peer setup. They all just connected to the office directly and worked off of the central network. No data needed to be copied back and forth. Remote staff got their work done with minimal technology issues. Partners and staff members in the office can view the audits in real time while the remote staff was working on the audit. Fewer issues and greater productivity was the result, plus the security of the data never leaving the office.

And the benefits didn't end with audits. Because partners and staff can now work from anywhere they can work with all the firm's tax, accounting, auditing, and billing applications from wherever they are, whether it be home, office, client, or anywhere else. Billing was more efficient, because staff did not have to wait until they got back to the office to enter their time spent at the client. They can enter it right on the spot at the remote site or even while working from home. Bad weather didn't stop tax accountants from continuing to crank out tax returns from the comfort of their homes, all while not losing valuable time during tax season. The efficiency and productivity benefits to CPA firms of remote application access software like Citrix and similar products are almost endless. As early as the 1990s, large pharmaceutical companies began using a form of remote access software known as field force automation for their widely dispersed sales staff.

Even the process of writing and editing this book has employed technology in the form of collaboration software programs such as Google® Docs. Musicians and filmmakers are bringing together remote participants across the planet to produce their end products using various collaboration platforms for audio and video production. Even video production projects have greatly benefited from the proliferation of remote productivity tools, with the videographer in one location and the presenter in a completely different location. For example, I had a video commercial done for my business. I never left the comfort of my home during the production. I worked out of my home office, using the fairly high resolution recording capacity of my phone to record my presentation, while the videographer, working out of her home in a completely different state, directed the production and received the raw recordings as they were being done on my end.

There is no doubt that businesses that take the "business investment" approach to technology see great benefits in helping their businesses to grow and drive more revenue. With the right technology in place, systems will be more reliable, enable greater efficiency and productivity, service clients/customers in a more optimal way, and ultimately grow in revenue and profitability, and potentially overtake the competition.

> *What technology are you using to enhance your business?*

Chapter 3
Cyber Security:
Expense or Investment?

When people think about cybersecurity, they're usually thinking about protecting themselves from the bad guys by making sure they have all the proper security protection tools like antivirus, firewalls, and other types of security tools to make sure that their business remains safe. In reality, there is a lot more to it than just that. What often gets lost in the mix is the impact to the business. Business owners view these initiatives typically as an expense, something that they must do — some will say it's like insurance; it's something they're paying for, and not receiving any real, tangible benefit from it for their business. It's just money out the window for something that they need to have to protect themselves.

Can you envision that spending money on cybersecurity can, in fact, help move your business forward? Perhaps even in a profitable, financial way? How you connect to this goes back to what I discussed regarding the cost-based view of business owners versus the thought that they're investing in their business. Of course I'm writing in relation to IT, but it affects all parts of business, even cybersecurity.

Two Scenarios

Let's look at a business that takes the cost-based view. They want to spend the minimum amount possible. They'll set up antivirus software on their computers and that will be that. Maybe they have a firewall. They have some backup software running. They're spending money on these tools. Okay, they say, "I have cybersecurity in place, I'm done." And they push it to the side, and they go about their business, minimizing their cost.

They feel that that's enough. The problem is that it's *not* enough. Antivirus and some other basic security tools and devices here and there is <u>not</u> cybersecurity. While antivirus and firewalls are important, they are not the entire strategy.

The issue is that with the cost-based view you are only doing a very small subset of what constitutes true cybersecurity. This is in stark contrast to the cybercriminal, who is typically highly motivated by financial gain. They will work to get the best tools and systems in place to maximize the return on their investment. You see, cybercriminals are in business, too. And their businesses are very likely far more profitable and have more financial resources at their disposal than the typical small business. The only difference is that what they are doing is illegal.

"A cybercriminal will do whatever it takes to extract the maximum amount possible from your business."

Because they are operating outside the bounds of the law, a cybercriminal has no limitations, and they will do whatever it takes to penetrate a business network, get to its important data, hold it for ransom, and extract the maximum amount possible from that business.

Now just imagine the business owner taking the cost-based view with all these basic security tools in place. They think they're safe enough. And they get hacked. Now they get hit with ransomware. They get a notification on their computers. They are caught with their pants down. Not only is their data not accessible, it's potentially compromised because the cybercriminals more than likely took a copy of that data for themselves, circumventing the backups that were in place.

Wait a minute... the cybercriminals got to the backups as well and deleted them! That was likely one of the first things they did. So now the business has no backups — all because there was not a proper cybersecurity strategy in place. The business owner wanted to save a few bucks. And now they don't know what to do because they didn't make a plan. Oops! They were more concerned about saving some money. Now, that all pales in comparison to what they will be losing as a result of the cyberattack.

All of this has very negative impacts for the business. The very basic thing is they can't access their data. But in the larger equation, operationally, they can't run their business. They can't service their clients/customers. Important data has been compromised, perhaps trade secrets, perhaps customer data, perhaps employee data. All of these have varying implications. Operationally, they are losing money. And now there may be legal ramifications as well. And if they are in a regulated industry, there may be regulatory ramifications. They have been brought to their knees and they may be compelled to pay that ransom even though that is not necessarily the recommended strategy. They may be left with no choice but to do so relying on the supposed promise of the cybercriminals to give them back their data and not have it posted out on the web for all to see. So, the minimal investment in cyber security that they made is now costing them dearly. Potentially the business may even shut down as a result.

"Cyber insurers have gotten smarter. If you did not have the proper strategies in place to protect your data, they may very well deny the claim."

In a best case scenario, the business may be out a significant sum of money. Oh, but you say what about cyber insurance? Will that cover the ransom payment? Cyber insurance is not a panacea and there are quite a few misconceptions about it. The cyber insurance may or may not cover the ransom. Cyber insurers nowadays have gotten smarter. They generally have restrictions. If you did not have the proper strategies in place to protect your data, they may very well deny the claim.

But let's say they did cover it. That doesn't change all the other ramifications to the business. Now the business owner may be tempted to sweep this under the rug and not let it be publicly known and move on. And that may work for a time. But, as with all things, secrets tend to get out there eventually. So, sooner or later the events of a cyberattack to their business will get out to the public.

And if they try to cover it up, that will get out too. So now they're in a double whammy. The news is out there now that they were the victim of the cyberattack, and that they covered it up. What would their clients/customers say about the fact that their business is covering things up — especially regarding the prized data clients/customers are trusting them with? They may end up losing their clients as a result, other potential clients/customers will likely look elsewhere to buy their products and services, not to mention all of the legal and compliance ramifications. Let's not forget about all the other stakeholders in the business. Like employees, for example, who may very well have insider knowledge about the attack because it was happening before their very eyes, and they saw the business owner try to cover it all up. What does that say to them? Not only may it compromise their company loyalty, but they may wonder how *their* confidential employee data is handled by the company.

This is all a domino effect stemming back from the decision to save some money on cybersecurity, thinking it was just another expense that needed to be minimized, with the assumption that having a few basic security tools in place was enough.

Turning the Tables on the Cybercriminals

Now, let's look at a business that takes the opposite point of view. They are looking at IT as an investment. And, by extension, they are looking at cybersecurity as an investment. So, yes, they may invest in many of the same tools. But they will also employ other more advanced tools, best practices, and strategies as well. Beyond that, they'll have ongoing monitoring, watching the network for any suspicious activity. They'll be making sure that their systems are patched and everything's up to date, both on all the computers as well as all network devices. They will be making ongoing evaluations of their environment to make sure that they have all their bases covered regarding cybersecurity. They will train their employees to avoid suspicious malware, such as phishing scams. Keep in mind that every connected device — even your printer — is hackable. They will have a full strategy in place, knowing what they will do, who will do it, and who they will call if an exploitation happens on their network (typically referred to as an Incident Response Plan). And they will practice the plan, much in the same way that football teams practice plays over and over again so they have it down cold when the real "event" happens. Certainly, they will have the proper cyber insurance in place, but they are not relying on it solely to pay the ransom. Cyber insurance has other resources available, including incident response services, legal assistance, and many others. All in all, it is a more comprehensive approach. Not just simple tools, but a whole strategy that involves technology, as well as people and processes. All this comes at a significant investment to the business on an ongoing basis, far greater than simply slapping on a few antivirus installations. However, rather than thinking of this as a cost, it's an investment in the business.

Why is it an investment? Why spend large sums of money on cybersecurity? Why would that be an investment in your business that will help to move it forward? Certainly, when you look at cybercriminals, they're determined to hack into a business and get to the important data and no business is entirely safe from that.

However, the proper cybersecurity approach is not the castle and moat approach. It's assuming that at some point you will get penetrated, and dealing with how to handle that is much like someone getting sick. If somebody gets sick and they have done all the proper health measures — exercise, drinking water, eating right, etc. — they may still get sick, but it will be mild. They may just get a head cold and will be okay in a couple of days. But if they did not do those things to protect themselves, they may wind up in the hospital, become severely ill, or even die. It is no different with cybersecurity. That is true cybersecurity. It's a more realistic approach. And why is this helpful? It assumes your network will be penetrated at some point. It does not necessarily mean that a breach has occurred.

One of your computers gets exploited, but, because you had the proper strategy in place, the exploitation gets caught early and the machine is isolated from the network. While it is a security incident, no breach occurred. The cybercriminals were stopped in their tracks before they got any further. This is where you want to be in terms of a cybersecurity strategy — ahead of the game!

This is where cybersecurity helps your business. You can demonstrate to not only your employees but also your customers/clients that you take cybersecurity very seriously by implementing a comprehensive cybersecurity strategy. Because we know that the cybercriminals are highly motivated to penetrate networks, we must have a plan in place to deal with them WHEN they get in — not if. We need to deal with that in a very efficient way to minimize the potential of a compromise to your data. And we need to be very transparent when this happens. If something happens on our network, we need to be upfront about what happened and what was compromised, if anything.

"When something does happen, we will act quickly and recover quickly."

It is a much more realistic, transparent, and honest approach. We're not saying that we're not going to be penetrated; we're saying that we have best practices in place and thus we'll be resilient when the exploitation happens such that the business will not be severely impacted. So, when something does happen, we will act quickly and recover quickly. What does this approach do for your business as compared to the other example, where the business did not want to spend a lot of money on cybersecurity?

The business that did make the investment will find itself in a far better position and their clients/customers and employees will have a lot more confidence in the business. Clients/Customers are more likely to do business with them because they know that the business is taking cybersecurity seriously. And the business may very well grow as a result. They may even be able to charge more money as opposed to the business that was cheap. While we cannot give you a specific ROI on an investment in cybersecurity, from the practical perspective, whom would you rather do business with? Would you rather do business with the company that doesn't care about cybersecurity or one that takes it very seriously and has dedicated the resources to protect their data? This Is why investing in cybersecurity will, in fact, likely help your business and help your business to grow and gain a competitive advantage over those that don't.

> *What approach do you take to cybersecurity in your business?*

Chapter 4
Managing IT:
CIO, IT, or CTO?

How a business manages its IT has a great deal to do with their management structure. There are several different roles in an organization that share responsibility for IT. The IT Director (or Director of IT), oversees the day-to-day operations of the IT department in an organization. The IT Director is not an executive level position. It reports to management, typically the Chief Information Officer (CIO). The IT Director is not concerned with how the IT affects the company on a strategic level, but rather maintenance of infrastructure, execution of IT projects, and support for end users, as well as supervision of IT staff. IT is a technical role, typically held by someone with a technical background.

The CIO, on the other hand, is more strategic. The CIO is an executive level position and oversees all IT operations in an organization and ensures the IT is serving the needs of the organization in the most optimal way. The role is a business role, not a technical role, and the individual filling the role does not necessarily have an IT background, though typically they may have an IT background even if their technical experience is many years out of date. They understand IT at a high level. The CIO is a part of the executive team and is involved in the planning of the organization. He or she is focused on the internal IT as a profit center for the organization, which may include increasing productivity and development of processes to drive more efficiency. A CIO is not to be confused with a Chief Technology Officer (CTO).

A CTO is more concerned with customer facing technology, while the CIO's focus is internal to the organization. In short, the CIO develops policy, while the IT Director is implementing policy. In a small business, the roles may not be quite as formalized. For example, a company may not necessarily have a CIO, and that responsibility will typically fall to the owner or partner or perhaps the CFO or COO, and they would oversee the IT Director or IT staff directly. In a smaller organization, the IT Director may very well have some planning and strategic responsibilities, but his or her responsibilities would likely still be mostly technical in nature. They may not have much IT staff and may be doing much of the technical work themselves, or they may not even be an IT Director and be more of an IT manager or systems administrator, which are typically more focused on the day-to-day IT support.

Because of the lack of a formal CIO role, the challenge is how to make sure the IT is being managed in a strategic way. The business owner or partner or CFO or COO is not equipped to effectively perform a CIO-type role. They are focused on running a business or some aspect of the business, not on how to strategically manage the business's IT assets. In this scenario, an IT Director with limited or no staff would be challenged to do anything more than putting out fires and servicing user requests, and not spend time on projects to improve infrastructure, not to mention cybersecurity, which is a full-time role in and of itself. The company may not even have an IT Director or even any IT staff, or, at best, they are low-level lightly experienced staff — or someone on their staff doing IT as a side note to their full-time role. Thus, it is the small business that is typically very challenged with managing IT as a strategic resource for the business.

So, what skill levels should these people have? An IT director' role, ideally, as we said, is at a technical level. I would characterize that as probably the highest-level technical overseer of the IT department. So aside from the essential technical knowledge, they should know how a network works, be familiar with various computer systems, and be experienced in those areas. They should know how to manage other people because they're managing an IT department. In a business with, say, 100 employees, there may be one or two other IT people. In a larger IT department, there may be one, two, three, four, or many more people who they're going to need to be able to manage. While being technically knowledgable is important, perhaps it is more important to have managerial skills and people skills — knowing how to deal with various behavioral types. In fact, those skills are more important than the technical because they don't have to be the smartest IT person in the world. In fact, no one knows everything about everything. They should be capable of leveraging IT staff or outside resources that may fill in the gaps in their skill level.

What about certifications? I think certifications could be useful. However, anybody could get a piece of paper and say they are certified. The IT director should have hands-on experience. That's more important than them taking a test. Just one last thing — they should be able to work with people at all levels of a company. That's very important.

What about a CIO? You might be wondering what sort of training experience or skills they should have. A CIO should be able to conceptualize the big picture since his or her role is not technical, their role is at the organizational level. They very much need people skills, dealing with their fellow executives, especially the CEO, to be able to conceptualize the needs of the IT department in the context of what's best for the business, and what will help move the business forward. They need to be the translator between the needs of the business, and what is needed from IT to help the business achieve its objectives. While they may not necessarily have current technical knowledge, they do need to understand IT at a high level. They don't need to know how to fix a computer. At a very high level they need to know what these things are: what a server is, or what a cloud is, for example, so they can plan that out with the IT team as the

CIO would need to have a handle on the general needs of the organization.

It is essential that CIOs have strong organizational skills; they need to know, from the standpoint of the IT side, what they need, and how to translate that to the business need they address, as well as to determine what is needed from a budgetary perspective — what projects are needed and when. They need to be able to communicate to the CEO and other executives that we need to do this project because this is how it's going to impact the business. They have to be able to translate "IT-speak" to executive team language, and vice versa. In the absence of having somebody with those skills on board, there may not be effective communication, because they may not communicate effectively with the IT director. You could have situations where you have these discussions and things may not be approved because management doesn't see how it's going to help the business.

What should your expectations be of the people in each of these roles? As far as the IT Director is concerned, they have to be very organized. They may be managing one or more people. Documentation and procedures are critical. The IT director and his or her staff needs to document everything, so that person knows what's going on and he or she can summarize it in reporting to the CIO (or an Owner or Partner). You can then see where the gaps might be, as well as be able to indicate what the IT department needs, and why they need it. Another crucial part of documentation is documentation of procedures and infrastructure. The important thing also is that the IT department should be effective regardless of who the individual people are in those roles. Procedures of how to do things in the IT department, along with the documentation on the infrastructure, will enable any tech slotted into an IT role to hit the ground running and not have to spend time re-inventing the wheel.

When a company hires for the IT Director role, they should have certain prerequisites. Certainly, they should be experienced technically, but do not necessarily need to be the smartest one technically. They need to be able to manage other people, communicate with higher-ups, and they also need to manage vendors. Often, since they can't know everything, they need to bring in a third party to help them implement a project. Remember, the IT Director is more at the operational level, as opposed to the CIO. The CIO is not getting his or her hands dirty. The IT Director must also be somewhat adept at budgeting. While a lot of that may be handled by the CIO, the IT director should know about those things as well in order to communicate with the CIO. If they feel they need to implement an infrastructure project, such as migrating to a cloud-based email system or migrating their physical servers to cloud-based virtual servers, they need to know how to budget — particularly if it's a small IT department or a small business where there may be only one or two other IT people and they may not have a CIO. Budgeting may fall in large part to them in those instances.

"The CIO brings conceptual, forward thinking into the future."

As to specific expectations of the CIO, leadership is important because the CIO leads the internal IT organization at the executive level. They can't talk down to their IT staff or IT director; they must have good interpersonal skills and be able to know how to manage people and how to get the information they need so that they can better inform management to help in the planning process. As I mentioned, the CIO has to look at the big picture. They need to have a sense of the business in its entirety — the purpose of the business, where they're going, and where they plan to be going. Then they need to be able to translate that into what is going to be needed from the IT department. Where is the business going to be in a year, two years, three years? They are paid to look at the long game for the business. They bring conceptual, forward thinking into the future, understanding what senior executives like to see. They need to know what the CEO and the rest of the executive team are looking for. They must be able to be a team player with them as an executive, so everybody has to be on the same page. Each of the executive team members, representing a functional area of the business, coordinates with the CIO, who represents the IT functional area. It's important to stress that it is not a lower function than, for example, finance; all parts are equal. Sometimes the CIO has to be firm in articulating how investing in IT initiatives is important to the business.

How should you measure the performance of your IT Director and your CIO? Starting with the IT Director, how you measure their performance in large part involves looking at what is the uptime. While the CIO technically is responsible for everything that happens in the IT operations, the IT Director is directly responsible for the day-to-day operations of the IT department. In the event of any failure of the company's core network, or core network systems, it's the IT director who is going to have to explain what happened and why it happened — and figure out how to minimize the chance of it happening again. They are managing the IT staff(or maybe doing the themselves) so they should be accountable operationally for what happens in every aspect of the IT department at the tactical level. You're looking at response time. They need to be very flexible in responding to any problems and be able to take those calls from the CIO or management if they have a question about something IT related. They must be very level headed. In terms of measuring their performance, uptime and minimized operational problems are most important. In a small business of say 100 or fewer employees, the IT director may be more accountable for all aspects of the IT operations and how it impacts the business, as well as cybersecurity projects, even incidents, though cybersecurity is a completely separate area of expertise. In short, the IT Director is generally left "holding the bag" in these micro-small businesses for any IT or security problems affecting the business, irrespective of where they originated from. Also, generally, IT directors are accountable for projects, making sure that projects that were approved by management move forward in a timely fashion and that they are operating within their budget.

When it comes to measuring the CIO's performance, they are responsible for making sure that the IT department is meeting the needs of the business and helping to move the business forward. They need to measure and test the results of the initiatives that were implemented and how that affected changes in business. They measure and test from a business level, how did this help the business? Did it achieve its intended objective? It's the CIO's responsibility to make sure that the IT department's initiatives were carried out and are serving the interests of the business. Measuring the result — whether it has achieved that objective, or maybe partially achieved it, or perhaps even not achieved it all, or even made things worse. The CIO is accountable for that aspect of it. If the project was completed and there were issues that significantly impacted the business, that's something where the CIO has to be held accountable, even though it is something that he or she may not be directly involved in, but the business budgeted for it. It's the CIOs problem when the project goes over budget and/or does not go as planned. The CIO is translating to management to get budget approval on projects for the IT department, and has to explain to the CEO why the project didn't get done or get done properly, or why it was not completed timely and/or over budget.

"The IT director needs to have knowledge of alternative technologies."

Let's take a look at the importance of continuing education for these people to pursue in order to remain relevant in a fast-changing world. From the IT director's side, they need to remain current so they should be taking ongoing training on various IT and cybersecurity topics. In reality what these IT directors do in smaller businesses may be very limited and focused so they're not getting a broad exposure to technology — only to the technology being used in that business. They need to be aware of and have some knowledge of alternative technologies. They need to maintain awareness and maybe search online to see what various other technologies are upcoming and maybe get hands-on training in-person and/or online. Certifications are helpful but they need to keep their finger on the pulse of what's out there because they are called upon to deploy projects and manage the IT infrastructure. It's a good idea to reach out and see what their peers are using. They should maintain a network of other colleagues, other professionals, as knowing and interacting with other people who have knowledge and experience is important. A good example of this is joining a peer group of other IT professionals. The IT Director has to know about what's out there, what's new, etc., in order to make good recommendations to the CIO/Management. They must be informed in order to articulate what is needed, as well as what may not be the right option.

Consider the potential impact of not being current, of not being on top of what's out there. They may experience operational problems, potential security problems, because maybe what they're using is not the best at that function. Maybe one particular vendor is better from a security standpoint, for example. Without that knowledge there's now going to be a security gap. Staying current will not only help with gaining knowledge on new technologies, but thoughts about what are best practices may come to light as well.

Do you always need to be working on the latest and greatest? Maybe not, because it's too new and has not been battle tested yet. The IT Director should have the ability to make recommendations on what to implement and what things should be avoided, and not jump the gun on the newest thing out there. For example, if the computer file storage system is not running optimally and needs to be replaced, it just may have reached its end of life. They're intrigued by the latest storage system on the market, which really seems to have some great innovative functions. It's very new, it's from a startup company. And the IT Director just fell in love with it, his engineer fell in love with it, so he recommended it to management and they implemented it. It turned out they spent months and months, maybe even up to a year, trying to get it to operate right in their environment. It caused great workflow issues, because the performance was so dreadfully slow. Staff and management complained to no end. Finally, they ripped out the solution and implemented a newer version of the product they had previously used, from an established vendor with a long track record, and it worked great. The cutting edge solution was not the right solution. It not only did not make things better for the company, it made things worse. It was, again, something very cutting edge and very new. It should have been avoided. And the IT Director should have known that. Very simply, don't fall in love with the newest thing that's out there; know about it, and maybe look at it again, perhaps in a year or two. See how it's doing in the marketplace. You need to learn about these new technologies, but don't jump on implementing them right away.

Part of it is maintaining awareness of the new technologies that are out there, maybe reading IT publications and doing some of your own searching on the internet, going to technology conferences, etc. Don't be persuaded by an overly aggressive and persuasive salesperson. Manage your vendors; be very strong about that. You need to have the courage of your convictions; you must be able to have that knowledge. Knowledge is power. IT Directors have to be knowledgeable about trends to a degree that they can be strong in who they deal with and who they don't work with.

MSP? YES!

In a small business with fewer than 50 employees, it will likely not be affordable to hire a full-time CIO or IT Director and the additional IT staff necessary to develop an IT strategy and execute on it. What is a small business to do? The answer lies in outsourcing to a managed services provider (MSP).

For a small business with no CIO and minimal or no IT staff, the MSP is the ideal solution. The MSP, at its core, serves as an outsourced IT department for the small business. It will perform the functions of the IT Director in overseeing the proper maintenance of the infrastructure, implement projects, perform ongoing monitoring, and interface with management on IT planning. Some MSPs have what is called a Virtual CIO, which serves to fill many of the roles a full-time CIO would perform. There are also Virtual CIOs that may be hired independently of an MSP, and the two would work together. Together, an MSP with a Virtual CIO will effectively give the small business a similar level of day-to-day and strategic IT management as that of larger businesses. How do you select the right MSP for your business? More on that in a later chapter... stay tuned!

How are you currently managing your IT?
Have you incorporated
strategic planning into it?

Chapter 5
Driving Business Results with the Strategic View of IT

When we examine how this strategic view of IT can drive business results, we see there are great advantages. First, we mentioned greater uptime and minimized disruption. Let's elaborate on that a little. By adopting the strategic view of IT, how do you achieve greater uptime and minimize disruption? To begin with, you're looking at this from the business operational perspective, not the technology perspective. You have to look at the workflow in the entire business from beginning to end, and how information flows through each part of the business. Look at what is working well, and also note what can be improved upon to increase efficiency and productivity in the underlying business operations. Another key area is to examine all potential areas where if something was disrupted or brought down, how it would impact revenue. With a strategic view of IT, it means taking a more holistic view of the entire business, then deriving a data flow diagram of how the data flows in the business and looking at the systems that support all those functions underneath. Then you determine whether that is the optimal configuration and if there are areas for improvement. In a business that has grown over time, typically each of the supporting systems were implemented without much thought as to how everything works together and if they are the optimal setting to support the continued growth of the business.

Perhaps, most importantly, you need to look at the areas which are driving the revenue in the business — all the dataflow areas, and you start with what's potentially the most important areas and look at the systems that are supporting them. Next, you upgrade or make changes to the underlying systems as needed. If there's a server, for example, that is supporting a specific data function, making sure that server is reliable, that it's up to date and has whatever it needs to get the job done. You determine whether it has enough capacity to run. By extension, you look at all the different points, the data points, and the systems that are supporting them, making sure that the underlying infrastructure is up for the job. You must determine that it's set up in a reliable way, with backup systems. That's more strategic, as opposed to just going in and looking at a server and fixing it. But more to the point, is there a better way for your systems to be structured that make your business run even better? It's not just a matter of accepting what is in place and fixing it or upgrading it. Look at whether it actually serves your business purposes and make changes to best align with those purposes.

To sum up, you have to look from the top down. Look at your business processes: assess how the underlying data is flowing, study the infrastructure underneath, prioritize by looking at what's most important in driving revenue, see what's supporting that, make sure that infrastructure supports what is needed and is set up in a reliable fashion and in a way that best supports your business processes. Then build on this going forward. Make sure that everything is accounted for, every potential data point in the business.

Let's take an example; suppose you had an e-commerce business. As we mentioned, regarding the workflow, there will be a variety of things happening — incoming orders, shipping, billing — all those various functions, and perhaps more. There are a lot of moving parts in an e-commerce business. You have your e-commerce website, your front end, and the interface that the customer interacts with. You have posted items up there with pictures of them, advertising what the items are, what the prices are, showing if they are in stock or not, and your cart for ordering them. That flows into the ordering part and needs to handle transactions occurring in a secure way, and the data is going from the order to whatever mechanism is connected to your inventory and physical items somewhere.

That system must talk to your inventory management system and determine which warehouse your items are located in and make sure it's selected correctly so it can be shipped out to the customer. There might be other elements between all those things, so there's quite a bit that goes on. While some of them may seem very minor, everything — every piece of it — needs to communicate with the other pieces, and it must be done flawlessly, without disruption. If at any point in that stream during the processing of that transaction something is not completed or somehow the inventory was incorrect and they don't have the item, and this was not communicated properly as it was not updated in their system, that can backfire. Then the company needs to backtrack to get it figured out. The result: the customer is displeased and may even go shop somewhere else. You can lose a customer very quickly in e-commerce. They are just a mouse click away from your competition.

"Once we've determined what we want to accomplish, then we build that vision."

Since there are a number of things that happen in a lead flow between the initial order and a customer receiving the item, how do you increase efficiency, using the strategic view of IT? When you take the strategic, more holistic view, first, know what you want to accomplish. What will assure that this system has what we need to drive that sale? What are the expectations from the customer? How do we make sure that the customer is getting what they ordered and in a timely fashion? Maybe we also need some mechanism for getting updates on the order, canceling the order, returning the order, etc. The bottom line: what exactly do we want to accomplish through this whole transaction? We need to decide all of this before we've even built the system or buy the software and other technology in setting up the website and other supporting systems. Once we've determined what we want to accomplish, then we build that vision.

So how do you increase the efficiency once you've built the system? That's a continual evaluation. Once you have the system in place, you're periodically evaluating if it's working as expected, or if there are deficiencies in the actual practice and how it's operating in a real-world scenario. This is not a one and done type of situation. You don't just throw up a technology solution and walk away from it thinking it's going to be working flawlessly and supporting your business in the way you expect it to. You must be continually evaluating it to understand where it is working correctly, where it is working wrong, and the areas needing improvement. Ask the question: what can we be doing better? You're always looking at how things are operating, if they're operating optimally, or if there are ways that we can improve how things are flowing to improve overall efficiency and productivity. It's a continuous evaluation to see where areas can be improved. In order to do that, you have to be building some feedback loops, determining where there should be some mechanisms for measuring and testing how the process is operating. That's something that needs to be built into the system.

"You have to adjust — continually monitor and adjust."

Productivity involves constant evaluation of how things are working now. If it's e-commerce, it is evaluating if the orders are being fulfilled and how quickly, and how well you are able to track the item ordered from the inventory selection to the shipments. If it's service based, productivity is based on how quickly you can get the client matter completed. You need to establish some benchmarks and expectations and then you have to adjust — continually monitor and adjust. You need to find where you are now, define a baseline, then implement the change, then see where you are after that. If it was a positive change, see what that level of change was. If it was a negative change, and made things worse, then determine why that happened. It is very likely something you may have overlooked in your assessment of the need and underlying technology elements. No assessment will be perfect, but if you did your due diligence in examining the existing systems, processes, and business objectives, and planned your technology project around these, most likely you will see improvement in the business versus what you invested. Accurately determining your baseline is key in order to test and measure. You want to know your return on investment. What was the increase in productivity? What does that mean for revenue? How much additional revenue did that generate for the business? Ideally, you want to know your baseline productivity and the revenue, versus the productivity/revenue after the implementation of the technology solution. See if there's a direct correlation with the technology investment.

"By increasing efficiency and productivity, ideally that will have some effect on profit."

This all leads towards, of course, to increasing profitability. How else can you use the IT infrastructure to increase profitability? By increasing efficiency and productivity, ideally that will have some effect on profit. Again, you look at your baseline — where you were from a productivity and efficiency perspective — and see if it is also where you were with your revenue, where you were for profit, at least in the particular area of the business that you are focused on, whatever product or service that you're working on to improve your revenue. Where are you now? What was the net increase in revenue? What is the net increase in profit from that one change or whatever IT initiative you put in place? You need to isolate what you're measuring. Make sure it is attributed to whatever product or service delivery you're looking to improve, or it could be what the overall change in the revenue and profit is for your business. To sum up: you need to know where you were before you made the change, then evaluate where you are now and see if there's a correlation from that change in the profit of the business as a result of making that change in your technology.

There are certain specific things that would generally result in increased profitability by making those efficiency and productivity improvements in your technology. A lot of it is going to depend upon activities that you're involved in. If, for example, you're a service-based business like an accounting firm, if you upgraded your infrastructure, now you're able to process tax returns faster and process more of them. Now you can service more clients and bill more and make more money as a result of that. If you're producing widgets, making improvements in your infrastructure like implementing, for example, an ERP system — now you're able to produce more widgets and make more money. If you're in the retail business, maybe better inventory control will allow you to know precisely how much of each item you have and their respective locations. When somebody places an order, you can redirect it to the closest warehouse, so orders get fulfilled faster. As a result, you will have better inventory management and know when you need to order more products. Orders get fulfilled faster and the customer has a great experience. That would likely lead to greater profitability for the business.

Often, one of the goals of a business includes increasing its ability to respond to markets quickly, changing direction, etc. By improving the technology, you can affect the ability of the business to turn on a dime. A lot of it is based on handling the information flow, having infrastructure to support that and having systems be able to talk to each other that need to. Going back to our retail example, you'll know exactly how much product there is in Warehouse A because all your systems are talking to each other and communicating properly. Making sure that your technology infrastructure is there to support those functions will enable the business to operate more quickly. You have access to information on the products you are offering and if you see there is excess inventory, you can make a decision on the spot to have a sale. Or if you see exactly how much a product is selling well in a given area, you can advertise somewhere else based on the real-time data. You are confident in doing so because you would know if it's doing well in one area with X amount of advertising, you can allocate more to advertise in another location in order to boost sales.

The key thing to remember is that, as a result of the investments you made in technology, you now have up-to-date, operational knowledge about what products are selling and not selling, allowing you to make informed decisions in a timely fashion on how to allocate resources for even greater profitability.

Service industries would benefit as well from this approach. With better automation, you can utilize your staff resources better and more efficiently. For example, in an accounting firm, partners and staff can process client information at the client site while collaborating with other people at the office. Instead of the partners and staff in the office waiting to get the information, they can just work online with the remote personnel at the same time. They'll be using their time very efficiently and the job will get done faster, and the firm will be able to take on more, similar jobs.

In terms of competitive advantages, using this strategic approach allows you to have information at your fingertips. In the case of the retail operation, they know their inventory, where the products are located, their sales levels, and where they are selling well and where they are not. That enables them to make better decisions. They know what marketing resources to allocate to which products, what they need in terms of warehouse space and where. Amazon has done a lot of that. They have warehouses across the country to enable them to fulfill delivery in a few days or even the same day.

Back in the 1970s, the greeting card business was very much ahead of the curve. They had all those slots with the cards on the shelf, and there was a certain amount of space allocated to each type of card by price and by genre — a mother birthday card, a father birthday card, a brother birthday card, a sister birthday card, at different price levels. Every time they reordered that was all being fed into a computer. At the end of a year, they did an analysis of how each store's space was allocated. If they saw that they weren't moving enough of a certain card, they would then shrink the space allocated to that item in the retail display and increase it for something that was selling more.

You may have seen that in your local supermarket. I sure have. All of a sudden, you don't see an item there anymore and they even took out the signage where it would be located on the shelf. How frustrating it was for me when I could not find my favorite oatmeal on the shelf anymore at my local supermarket! And I went there week after week and sure enough it was not there. Even the signage for it was gone. I could not even find it at another of the chain's stores! The large retail firms like Walgreens®, CVS®, or Costco® use what they call velocity reports, where they track how many times the product turns over. If it doesn't reach a desired velocity, then they will throw it off the shelf. Apparently, my local supermarket chain does the same thing!

"Technology investment essentially allowed Amazon to become the dominant player in its industry."

Amazon is a perfect example of a company that's reaped great benefits by investing in technology, growing by leaps and bounds over their competition. They've continuously invested in technology, even at the expense of profit. Their margins were always very slim, but they just continued pouring money into technology and they muscled through and largely took over as far as e-commerce is concerned. There were a number of competitor e-commerce sites that did not survive the onslaught by Amazon. Either they were put out of business by Amazon or Amazon bought them out. Amazon's huge technology investment essentially allowed them to become the dominant player in the e-commerce industry. They're the epitome of a company that made the investment in technology and achieved game-changing business results.

Other industries have adopted this strategic view of technology to varying degrees as well, which has allowed them to progress. Look at how the automotive business had to reinvent itself in a lot of ways, technology being one of them. Investment in automation in their factories enabled them to manufacture with more precision. That again speaks to the quality that allows the production team to move quicker, better, with fewer defects. The machines don't take breaks and don't make mistakes (if they are programmed correctly, but then again that is a human mistake!). Any type of manufacturing environment would benefit from that.

A technology that is typically very impactful on a business's bottom line is Customer Relationship Management

(CRM) software. John Hauryluke, owner of CRM Cloud People, has been working in the CRM field for over 20 years, performing CRM design and implementation work for small and mid-size businesses. John talks about a few of the clients he has impacted over the years.

One such client was a telephony company based in Florida that provided services to businesses across the country. They were using a CRM system that was not up to par for what they needed for telemarketing. They decided to move to SalesForce and called John's group in to help them. John put together a strategy of exactly how they would build out the system. The key aspect of it was integration of call telephony into the SalesForce system. Once the implementation was done, the client hired 20 additional people to the three they already had doing outbound calling on a day-to-day basis using the system. As a result, the client's bottom line increased by tenfold, simply by implementing SalesForce and integrating it with telephony.

Another client John worked with was a technology services company in New Jersey that was using SalesForce, but did not fully leverage its capabilities, particularly in the area of reporting and integration with all the operational areas of the business. For example, SalesForce didn't integrate with their finance and inventory management system. Hardware inventory was tracked on a legacy application, which was totally separate from SalesForce. Customer contact and account records were also on separate systems as well. John's group uploaded thousands of account and contact records, brought in the inventory control, and unified all of the operational areas within SalesForce. This allowed management to gain accurate insight into the activities of the company and make more informed decisions. Customer account tracking was more precise and streamlined, allowing account reps to know when contracts were expiring, what services were being provided etc., which allowed for a more efficient sales and account management process.

> *What is holding you back from investing in technology to help achieve better business results?*

Chapter 6
Preparing for Cyber Security as a Business Technology Strategy

In previous chapters we discussed the potential impact of cybercrime on your business. We will now demonstrate the best ways to proactively protect your business.

Previously we talked about proper cyber security preparation as a business technology strategy and why it's important for the financial results of the business, just like any other technology-related investment. Now let's get into the finer points of cybersecurity preparedness. When it comes to cybersecurity preparedness, the first thing to know is that it's not an "if," it's a "when." At some point, every business will experience an attack of some kind. Let's start by talking about how a cyberattack can disrupt or destroy a business. A cyberattack can potentially be very devastating for a business. It could be anywhere from a minor nuisance to the ending, where a business just shuts its doors — and anything in between.

Let's discuss some of the typical disruptions and what kind of systems are disrupted. To begin with, the modus operandi of threat actors are to get to the business's important data. They use various means to accomplish that. The first thing they need to do is get a hold of the network. Typically, that is done through a phishing campaign, trying to trick you or one of your employees to take an action through a phishing email and compromise your systems. Once they get into one system, they need to figure out where they are, how many privileges the person has and, basically, bounce from system to system expanding the reach within the network until they get to the good stuff: your business data!

Once the threat actors get to the data, what do they want to do with it? Make it unusable for the business so that they are compelled to pay a ransom to get it back. That is the threat actors' demand. The other thing that they'll usually do with it is take a copy of that data for themselves, meaning that they will extract it, then upload the data to their servers that they control. In combination with encrypting the data and making it unusable, they put a further stranglehold on the business. So, even if a business has proper backups done the right way, they may still end up paying the ransom. The threat actors have their own copy and may threaten to put it out there on the web, exposing potentially very compromising data, which will negatively impact the business. Potentially, if they do not get paid the ransom, they're going to do whatever they want with it. It is going to go out on the dark web, posting it publicly and/or selling it to the highest bidder, who will use it for their own nefarious purposes. The dark web is just loaded with places where threat actors sell information they took from their various ransomware expeditions.

We mentioned that phishing is one way threat actors gain entrée. Let's look at some of the other ways that the threat actors may infiltrate your system. One major way is to exploit vulnerabilities in the IT infrastructure, such as computers and/or servers that are not up to date with patches, or network equipment that may not have the proper updates to their firmware. In short, potentially every network device on the network may have a vulnerability that was not patched. It is not just the computers themselves, it's not just the servers. It is every device that's on your network that could be an entry point. Think about it. Even your printers have firmware, your copy machines, and also wireless access points. IoT devices are a big thing now, such as smart home automation that controls your appliances. Even they can have vulnerabilities that threat actors can use to compromise your network. Anything that is connected to the network needs to be properly managed and updated. Your computer system's vulnerabilities are a big factor, including the operating system (eg. Windows, Mac, Linux), but keep in mind that you need to be mindful of the application software you use that is installed on your systems. That is a challenge because you're involved with third party software developers you bought the application software from. The onus is on them to make sure that the software they are supplying you with is up to date. If it is not, that could be an entry point. More often than not, security takes a back seat with application software developers who are more concerned with adding features and bringing their software quickly to market.

Applications (apps) are increasingly now in the cloud, and thus another area of exposure. Cloud apps are no different from traditional on-premises software in that they can have vulnerabilities as well. It is exacerbated by the fact that the cloud-based app is typically publicly accessible via the web. If there is a vulnerability in the web infrastructure and/or the app, you have the potential for a major compromise. Additionally, you have no control over the app vendor. You are trusting that they will do the right thing in terms of cybersecurity best practices. If you are using an application and threat actors compromised it, they potentially could get access to your information inside the app.

Obviously, you want to minimize the impact of a cyberattack when it occurs. Since it is potentially inevitable, let us look at ways a business can minimize the impact of an attack. Let's start with the basic premise that sooner or later, you're going to get hit so it's not a question of the traditional castle and moat approach. That is going to lead you to failure because you could fortify all the edges of your network and all the threat actors will do is just bypass them and get inside your walls. Once inside, there is likely minimal protection if you have not planned properly. There must be a defense at multiple levels. In addition to the outer edges of your network, you must have protections running inside the network as well. You need to have people who are actively monitoring your network for signs of an intrusion. Once the threat actors have compromised a user's computer on your network, they have compromised the inside of your network. Thus, all your individual computer systems and your network infrastructure need to be monitored internally. In addition, you must also log on your network and there needs to be security people on the other side of that to review the logs because the threat actors are going to try to keep themselves as quiet as possible. With logging and security people reviewing those logs, if something is abnormal, the security people would have the ability to take action and isolate the affected system — which may be one course of action that can potentially save your company from something far worse.

"The biggest exposure is you and your employees."

Part of protection involves your employees and you as well. Management needs to buy into a cybersecurity mindset and security awareness mindset. Ironically, the biggest exposure is you/management and your employees, which means continual education is key. As far as security awareness, this involves impressing on both management and the employees, even in the thick of their workday, that they still need to keep cyber awareness in the back of their mind, always. What kinds of things should personnel be aware of? They should be aware of the sheer cleverness of phishing emails. For example, somebody could email you saying, here's my invoice, it's overdue. You take an action to open the file to pay it, and that invoice might be filled with malware. Case in point, they may have hacked into one of your vendor's email systems. As a result, the invoice never came from the vendor; it came from the cybercriminal that hacked into *their* email system. The best advice: if you get something that is unsolicited, question it. If somebody sent you something, a link, or attachments, or anything of that nature, and you were not expecting it, then you need to step back and contact the person that supposedly sent it to you, not using any contact information in that email. Instead, use the contact information you have on file to contact that person. If you are in doubt, always call the person, again, using the information you have on file. Alway be aware of the possibility that it is a phishing email.

You also need to scrutinize very carefully the email address and see if it is indeed the real email address of the sender you think it is from. Simply put, study the email carefully, because very often it is an impersonated email address that isn't exactly correct, but may be similar to the address of your contact, or it could be totally different. It could very well come from a domain that is worded very closely to the domain of the legitimate party — such as infinity.com may appear as infinlty.com (do you notice the "l" instead of an "i" in the second version?). Very often, they look very, very, very legitimate. The threat actors have become extremely proficient in creating a genuine looking email. It may look like they are from, say, Amazon, or one of your vendors. It is not really from them; it's from somewhere else. It may tell you to click on "this" to get your gift card or something of that nature. Just be aware that looks can be deceiving. Scrutinize the email and scrutinize the links. Are the links valid links that look like links from the real company? Or is there something different about them? You must be very mindful of that.

"Conduct practice sessions for incident response with all parties involved."

In regard to the question of when you must be prepared for what happens, it is important to note that everybody in the company, from the executive team all the way down to the employees, need to be vigilant. You need to know who will be involved in the response, determine who will orchestrate the response and the team members executing on the response, and train for that. Conduct practice sessions with all parties involved: who will coordinate the incident response, what management and employees are going to be involved, even third parties. One thing to keep in mind is that it is not necessarily a given for the business owner or CEO to lead the incident response. That role should fall to someone who has the strength and temperament to lead the response forward under highly stressful conditions. You must know exactly what you are going to do when it happens and who will do it. When it does happen is not the time to be figuring out who will be involved and practicing the response because that will inevitably lead to disaster.

How can businesses minimize the impact of a cyberattack? It is a multi-layered process. Part of it is to have the right technology in place, the proper security tools in place — that's a given. Also, again, having the right people in place to monitor your network, processes to orchestrate the response and deal with any potential incidents, as well as good employee (and management!) awareness. Knowing how you are going to respond if something happens is important, because the more prepared you are, the more likely you are to have a favorable outcome. It is multifaceted with many different levels. It is not just a one-and-done. It is a continuous process, much like with our own personal health. You need to eat right, to exercise, and to take your vitamins, so if you catch something, you will get better more quickly. The impact of the illness is probably going to be much less than if you weren't taking care of your health. It is the same thing with cybersecurity. It is no different for a business. Engage in proper cyber hygiene, make sure we have the people, processes, and technology in place, and have it all well-coordinated. If you get hit with a cyberattack, the impact will be much less, and you will be able to absorb the impact and continue to run the business in an operational way.

Even though it is possible that it might not be at full capacity, depending upon the severity of the attack, proper preparedness will help because you have planned for it and trained for it. Then you bounce back quickly and continue to operate, moving forward to continue the business operation rather than being held back by trying to recover from an attack that brought your business to its knees.

By now, I hope it is evident that you need to adjust your business strategy to address cybersecurity. To reiterate, the business culture needs to be more security-aware. This needs to be part of what they do as a business — not that it is going to be at the exclusion of all else, but it just needs to be incorporated into all the business processes. Everything. If somebody's working day to day on a project, they still must have some awareness in the back of their mind. If they are on a deadline to finish something — perhaps they are an accountant preparing a tax return on a deadline — in a rush. Having that security culture in mind helps to keep an awareness not to click that phishing link and compromise the system. In the heat of the moment, it is easy to forget. But, if you have maintained that security awareness, it is less likely to happen. And make sure you have multifactor authentication enabled wherever possible, particularly on cloud-based applications. This is one of the simplest ways to help protect your business, adding one or more additional authentication sources beyond just a username and password. A couple of examples are a mobile phone, which receives a code by text or call, and an authenticator app, which is installed on your mobile phone and generates a code. Whatever method you choose, typically the authentication code that is generated would be typed into a field in the application/device logon screen in addition to your username and password. There are other types of authentication applications and methods as well, so the procedure could vary. Overall, some are more secure than others, with text likely being the least secure.

If you do suffer a cyberattack, it is important to learn from the experience. Once your business is attacked, and you have responded to the incident, what can you do to further protect yourself and your digital assets against future attacks? Just know that the incident you responded to is not going to be the end of it. To be clear, a cyber incident doesn't necessarily mean you were breached. You could have a cyber incident and your data was not compromised. The intruders may have just touched your network. If you isolated the problem, you may have prevented more severe actions. In the opposite scenario, you had a cyber incident, they got your data, and you responded using your incident response procedures. After it is all over, once you have restored the business back to where it was before, depending on what happened, there will be a learning phase. You need to learn what happened. Why did it happen? How do we fix it for the future? Now, if you are doing the proper cyber preparedness, this is part of your process.

Another important component in your preparedness involves having cyber liability insurance. Why is it important to have this coverage and what will the insurance carrier expect of your company? When this insurance first came out, it was relatively easy to buy; check a few boxes and get the insurance. Now, the cyber insurance carriers have become far more diligent at scrutinizing your business. Generally, now, if you can get a policy, you are going to need to implement many of the strategies that we recommend — you must do nearly all the things I'm telling you to do to get the policy. The insurance companies are looking to minimize their risk because they've been burned very badly due to all the ransomware attacks and they have had to pay out quite a bit. This is why they are now checking to make sure companies know how to be cybersecurity aware and have created more stringent requirements to even get the insurance. As we indicated, having cyber liability insurance is just part of the strategy — not that you want to rely on it. It's just one tool. Keep in mind that it's not just about having the ability to pay the ransom. That's just one aspect of what the insurance carriers do. For example, they provide a breach attorney, they will provide incident response, they may provide a call center for fielding calls from customers and other interested parties, and many more things that the cyber carrier has under their umbrella that you would want at your disposal as part of your overall cybersecurity strategy.

Assuming that you have complied with all the requirements, you received the policy, and are then faced with an incident, how do you respond and get the insurance carrier involved? In general, what does the carrier expect of the insured company? They are expecting that the company employs a proper cybersecurity strategy for their business and maintains it. They must have some assurance that you are actually maintaining what you have put forth that your company is actually doing, whether those are strategies you employ internally or outsource to outside providers to implement. They will also be looking for you to show evidence that you have trained your employees properly, that you are performing ongoing monitoring, that you have all your software and systems patched, that you have checked for vulnerabilities, and a whole host of other cybersecurity best practices. If you get hit and the insurance carrier discovers that you said you were doing a particular protective action, and you got hit because you weren't doing that, they are not going to pay out on your claim. You will be on the hook for the cost of the damages to your system and your business, and whatever other repercussions that may have occurred.

In keeping with the theme of this book, meaning the importance of investing in technology and viewing it as a profit center, let's look at cybersecurity in that light as an investment. Business owners need to realize that this is a must-have for sure. Cybersecurity is something you should not think of as an expense; it is actually an investment in your business. Sure, even with that investment, you could potentially still get hit, but because you have a sound strategy in place, you're resilient, so the impact may only be minor. Also, keep in mind your competition — they may or may not be doing it. You want to make sure you are at the forefront of this. If you take on the view that it is just another expense, you likely minimize your costs and then when you get hit, the stakeholders, like your employees and your customers/clients, may desert you — you may even lose clients to your competitors as a result.

You want to demonstrate that you are making the investment and doing all the right things and can point to a culture of security awareness so that it gives confidence to not only your employees, but also to everybody who does business with your company — your vendors, your clients/customers, and others who may consider doing business with you. They will know that your business is a place your clients/customers can trust with their private information.

Consider that there is a lot of sensitive information that is entrusted to businesses. Therefore, security becomes extremely important. Whether you are a professional service firm, like an accounting firm that has your clients' social security numbers, or you are a medical office or hospital that has personal health information, or an attorney with sensitive legal information, or if you're an e-commerce company, and you have customer's contact and payment information, people are not going to have confidence in your company if you can't demonstrate that you are employing cybersecurity all around. The key is, if you do it right, it will be something that can help your business to grow because you are demonstrating cyber resilience strategies. If you are showing that you do all the right things with cybersecurity, you are doing everything that you should be doing in a multi-faceted way, you look stronger versus your competitors that may not be employing these strategies and this will only benefit your business in the long run.

Let's look at it from another standpoint. Supposing your company is considering getting outside investment. Wouldn't that cybersecurity practice also help to give the investors and/or the lenders a comfort level? Absolutely. Investors and lenders want to know that such business risks are addressed. Cybersecurity is an important business risk. One of the things potential funders or lenders most likely look at is your cybersecurity strategy to ensure that if they either bought your company, or provide any level of investment or funding, that your company is doing the right things. They want to make sure their interests are protected. They may very well strongarm your company into making the proper cyber security investments before they provide the investment. They may very well also clean house if you haven't been engaging in proper cybersecurity practices.

Back in 2013, Blue Cross® and Blue Shield® of New Jersey had laptops stolen out of their offices and they had to give all the policyholders credit monitoring service for a year. They had to draft and send letters to every policyholder to make them aware of it — and obviously deal with the media on the issue. It just goes to show you it does not matter how big or small you are, any company could get hit.

Look at Home Depot®, for example.[1] You would think a company the size of Home Depot would have the best cyber security strategies in place. The caveat here though is the intruders came in through a back-handed way — their vendors. Credentials were stolen from third party vendors and once the threat actors had a foothold on Home Depot's network, they used hacking techniques to gain administrator level privileges and proceeded to move laterally on the network searching for the important data and/or systems. They eventually found what they were looking for: the POS system. They infected the POS system with malware, which resulted in customer credit card information being exfiltrated to the threat actor's servers.

1. https://www.arctitan.com/blog/case-study-data-breach-cost-home-depot-179-million/

The attack went undetected for about five months, from April 2014 to September 2014, before it was finally removed from Home Depot's network. The impact was the compromise of 56 million payment card numbers and 53 million email addresses,[2] with a total cost of $179 million which may include additional legal fees and settlements. But let's not forget about the reputational damage that was done with customers who may never shop there again, and even customers who were not affected may be reluctant to shop there in the future.

In a study by PCI Pal, 83% of consumers say they will stop spending at a business for several months in the immediate aftermath of a security breach, with 21% of those claiming they will never shop again at a business after it has been breached.[3]

With annual revenue of $151 billion in 2022,[4] the cost of around $200 million seems like a drop in the bucket for Home Depot. They seem to be weathering the cyberattack

2. https://www.upguard.com/blog/biggest-data-breaches-us
3. https://www.businesswire.com/news/home/20190917005012/en/New-Global-Research-Shows-Poor-Data-Security
4. https://www.macrotrends.net/stocks/charts/HD/home-depot/revenue

just fine. One can only speculate as to the impact of how many customers they may have lost because of the breach, either temporarily or permanently.

Since the breach, Home Depot has made additional investments in their cybersecurity and hired a private cybersecurity company to help them with securing their systems.[5] As in many cases with a cyber incident, it is ironic that a breach had to happen to persuade management to make the additional necessary cybersecurity investments. You would think with their resources they would have done this right from the start and minimized the potential impact. Wouldn't it have made good business sense to do so?

Uber

Uber was hit by a cyberattack on several different occasions, 2014, 2016, and two more in 2022.[6,7] In the 2014 incident, over 100,000 driver records with sensitive data were exposed. The threat actor was able to get access to the data through code

5. https://www.idstrong.com/sentinel/home-depot-could-data-breach/
6. https://firewalltimes.com/uber-data-breach-timeline/
7. https://www.ftc.gov/news-events/news/press-releases/2017/08/uber-settles-ftc-allegations-it-made-deceptive-privacy-data-security-claims

shared on GitHub (a cloud-based software development platform) by an Uber software engineer, which allowed access to the data stored in plain text on an AWS (Amazon Web Services) server. It is a common theme of the lack of security awareness in how software engineers handle writing code. The focus tends to be on building the feature set of the application and getting it done quickly in order to get the application or application update to market in the shortest amount of time. It goes back to building a culture of security in the organization and embedding security into all aspects of a company's processes. Security should be built into the code writing process itself and not as an afterthought. Plus, basic security practices such as multi-factor authentication were not implemented, which is especially important for cloud-based applications.

In this example, procedures around handling code and implementing multi-factor authentication would have helped to minimize the chance of a successful cyberattack. Moreover, the fact that future attacks have occurred means that Uber likely did not fully follow recovery procedures by learning from the incident, building tighter processes, or enhancing their cybersecurity strategy.

In 2016, Uber suffered a second cyberattack. Threat actors exploited a vulnerability and compromised data relating to 57 million Uber customers. Uber chose to pay the threat actors $100,000 to delete the exposed data. The fact that Uber was compromised again after about two years indicates they had not fully thought out their cybersecurity strategy. It was almost as if they were caught off guard and their reaction was to keep the matter quiet and try to protect their reputation rather than be concerned about doing the proper incident response procedures and properly disclosing the incident. In the end, their actions worked against them. The incident was exposed in November 2017 in a Bloomberg® report of the incident. Uber ended up paying $148 million in penalties in a settlement with the FTC over concealing the incident.

And there is more. In August 2020, cybersecurity firm Cyble® uncovered information related to Uber Eats customers and drivers found on the dark web — 579 customer files and 100 Uber drivers. Then in 2022 Uber was compromised yet again, with one security researcher in *The New York Times* saying it was "a total breach." The hacker used social engineering techniques, spamming Uber employees with multi-factor authentication requests until they finally granted permission to access Uber's systems.

Going back to our theme, had Uber made the proper investments in a cybersecurity strategy and promoted a culture of security right from the start, they would not have suffered the severe reputational damage done because of these multiple cyber incidents. And to have been hacked on numerous occasions should give a customer some pause about using their app and entering their credit card information, for example. Drivers may also have trepidation. Certainly, it has had that effect for me. I considered using Uber on multiple occasions. The one time I did, several months later I found that someone else was using my credit card number. Whether the two are related one can only speculate. But knowing about all the cyber incidents and that Uber's security strategy seems to have more holes than Swiss cheese, I would be reluctant to use them again.

No business wants to take on additional expenses, but the fact that more threat actors are coming onto the scene by the day and becoming increasingly more sophisticated, with seemingly unlimited resources at their disposal, it is not only your company's bottom line at stake, but also your company's public reputation, particularly with your customers/clients and other stakeholders, that make it essential to have a cyber security strategy in place.

> *Does your business have a culture of cybersecurity awareness?*

Chapter 7
Ramping Up Your Productivity with Software and Mobile Apps

When it comes to productivity software apps, you may ask yourself whether you really need them and what their value is to your business. When you analyze them from a cost benefit basis, can they save you money? Can they make you money? Let's talk about some of those apps. In general, productivity software and apps, if used correctly, will enhance your business. They will make you more efficient, make you more productive, and ultimately lead to more profitability. However, they can be used the wrong way, too.

If you go out and buy software programs haphazardly, without a thought as to how they may best work together, you end up with a hodgepodge of different apps which have no correlation to each other. You may in fact be less efficient as a result. To avoid that possibility, should they be bought as a suite? The ideal is to buy apps as a suite if you can because if you're buying it from one developer, all the apps will talk to each other very seamlessly, and will be very efficient in the way that your work flows. Of course, not every software application is offered as part of a suite. A vendor may just have that one app you think is good for your use and you buy another app from a different developer for whatever function that app performs. Sometimes you may not want the whole suite because the individual apps in the suite may not do exactly what you need. In the interest of efficiency, that's something that you must evaluate.

With that said, when you purchase apps, how do you know whether they will work together or not? If you buy them as part of a suite, generally they'll integrate well because if the apps are all from one developer, you are assured that all their apps talk to each other generally speaking. Conversely, if you are buying the apps individually from different vendors, often one software app can integrate with other apps, even if it's not part of a suite from the same developer, but that is not a given. You must check for integrations with other apps to see if the app you're contemplating buying is able to integrate with the other apps that you own or that you're looking to buy as well. So, there is some necessary homework to see what integrates with what. That may guide your decision on what to buy. So, how would you find that out?

In medicine, you can open the Physician's Desk Reference and you can find out what drugs interact with others. When it comes to software apps, the first place to go is to the vendor's website. If they are diligent, they will show what their app integrates with because it's in their best interest. They want to make sure that they are integrating with generally all the major other complementary apps in the industry. On the vendor's website, look in the list of specs for the application. If integrations are not shown, you may need to call them and find out from them directly. If they don't show it on their website, be a little skeptical. Every vendor should be open about what they integrate with. They should be open to integrating with complementary apps, even if they're from competitors. It shows that they are more open about how they do business. You don't want one that is closed and that doesn't offer any integrations because then you are locked into their family of apps.

Let's address, in general, some must-have productivity apps. While much of it depends upon the industry you are in, in general for most office type environments, you want the basics — spreadsheet and word processing, email and presentations. In today's business environment, consider adding communications type applications like video conferencing. That has become very commonplace now; Zoom®, for example, is probably the one best example — or Microsoft Teams which is offered as part of their Microsoft 365 suite of apps.

Let's look at a few specific industries: retail, manufacturing, and e-commerce. For retail, in general, regardless of what they are selling, retailers will need some kind of point-of-sale application. Whether they have a restaurant or are selling clothing or whatever kind of retail establishment with a physical location, they need some kind of point-of-sale system. That is a fundamental requirement for any type of retail. Depending on the size and scale of the operation, they may also need an inventory control system — particularly if they do a very large volume. If the business is involved in e-commerce, then it will most certainly need some inventory control system in place to track the inventory's location and delivery.

As for manufacturing, in general, manufacturers need to produce products very efficiently, in the shortest amount of time possible. To distinguish from information technology, there are operational technologies that are applied to manufacturing — systems that will manage the manufacturing process, and there are applications that may interface with or control those technologies.

In the architectural sector, a business would probably need a Computer Aided Design (CAD) program. This would be a must have for an architectural firm, engineering, or even electrical contracting for example. AutoCAD® is a leader in the architectural and engineering industries. The CAD application makes it so much more efficient to create building designs with great precision and efficiency. There are also applications that will scan entire buildings.

"Wait at least six months for any new software application for the bugs to be worked out before purchasing it."

Oftentimes we hear people say that they don't want to buy software that was just released because there may be bugs that need to be ironed out. You may be wondering about how soon you should embrace a particular software application or a new major release version of it. For any new software application or major release that comes out, I generally recommend waiting at least six months for the bugs to be worked out. For the larger, more complex application, I might say wait longer — perhaps a year. This would hold true regardless of whether it is off the shelf software or cloud-based applications. Software is software. Either you are installing it on your systems or someone else is managing it in their datacenter and offering it to you as a subscription-based cloud service, more commonly known as SaaS (software as a service).

On the other side of the coin, you might wonder how long a software application is still viable before it needs to be upgraded or replaced. In part, it is going to depend upon the developer — whether they are actively supporting the product and the likelihood that they will continue to do so for the foreseeable future. If they are doing a good job in developing and maintaining their software, making sure that it's properly updated on a periodic basis, then it might be something that you can retain for the long term. If the developer is going out of business and it's just not being supported anymore, then you might need to think about some alternative program. This is something to consider particularly if you have had an application for a long time and it is a core part of your business process. If you see that the developer is looking to fold, then you need to figure out what you are going to do once the developer closes its doors. If it is a piece of software that runs on your computer systems, assuming it will still run with or without the developer, how long will it run for? There may be glitches or compatibility problems with newer operating systems. While it may be a significant amount of time and expense to migrate, it may be prudent to plan for it as it may cost you more down the road.

If it is a SaaS application, it is even more critical as the SaaS app developer is also hosting the application, so the app will simply be shut off when they close. Ideally the failing developer will be acquired by another developer, though the caveat here is that the acquirer may not intend to continue the app for the long term, but rather try to steer you to their comparable app, which may or may not fit your needs. These are major reasons why, if you are evaluating software applications (on-premises and SaaS), to also evaluate the long-term viability of the developer behind the application. Are they growing? Are they a startup or are they more established? How are they funded? How financially stable are they? Do they have a well thought out business plan? Does the executive team seem forward thinking and confident? Where is the industry going? What are the trends? How quickly do they deal with bugs in the software, feature releases? Security updates, cloud application security? How well are they maintaining the application? These are all things you need to consider before investing in an application.

Today's Business is Highly Mobile and More Adaptable

There are numerous mobile technologies that a business can leverage to its advantage. For example, you can take advantage of VoIP (Voice over Internet Protocol) and make use of it in your everyday business for greater productivity. One of the greatest benefits of VoIP is that it allows you to have a virtual office. You can have your phone at your office or you can take the phone home and you can answer it from there as if you are in the office. Or you can make or answer calls from an app on your smartphone from anywhere. It makes the physical office less relevant. It fits very well into our society where more and more businesses are scaling back their physical office space, saving a lot of money on rent and all the other overhead of maintaining a physical office, with increasing numbers of remote employees. Many businesses are getting out of offices entirely and going all-remote. VoIP is a big part of enabling that because now you can call a business and the call can be received from anywhere with an internet connection. Furthermore, VoIP has also reduced the need for a lot of extra expensive hardware. Remember, not that long ago, you had very expensive phone systems with physical, corded equipment. Again, you were locked into having a physical, stable space. Those also tended to be very expensive to install and they required proper maintenance. VoIP eliminated all that. You don't have to have large, expensive phone equipment boxes in a wiring closet anymore. It's very efficient and typically only requires an internet connection. You don't even need a physical hardwired phone anymore. Your smartphone can become your office phone or you can use your computer to make and receive calls.

Part of the transformation of businesses into more mobility includes video conferencing. In the past, this required a whole different scenario. Today, it's simpler, faster, and less expensive. Video conferencing has greatly enabled the virtual world of mobility today. In earlier days, if you needed to have a meeting or any kind of a conference, you had to have people traveling and coming into a physical place to meet. If you wanted to meet with someone, they would have to come to your office or you would go to theirs, or meet over coffee or a meal somewhere — and you'd have to figure out all the logistics of doing that. That's all gone away with the increasing use today of video conferencing. Now, if you have company meetings, you could literally have a meeting with employees all over the country, all in one session. Your employees can be anywhere. This has created a whole new paradigm of how a company is run. If you want to have a meeting with business associates or a client, you can do it totally virtually. You don't have to go to their office, you don't have to have lunch somewhere in a noisy, crowded restaurant, you could just do it as a video conference. No need to figure out where to park to meet over coffee! Just imagine the implications of that because you can literally sit at your desk at home all day and have video conferencing calls all day. You would never be able to achieve that without current video conferencing technology. Consider that, prior to this, conferencing required you to go to a conference center or hotel or someplace where you had to rent a studio type environment by the hour, or where you might have needed a satellite transmission. That was much more costly. Today's video conferencing technology has eliminated the logistics involved with getting people to a space, as well as the rent and the overhead. With a "desktop" video conference you can just have a webinar. There's no room to rent, you don't have to provide lunch or worry about travel. You can just sit at a desk and with a few mouse clicks attendees are at your conference; you can even do it from your mobile device.

Video conferencing is just the start. Common office productivity tools like word processing and spreadsheets have been brought to the cloud. The best examples of these are Microsoft 365® and Google Workspace®. In these products, the essential tools are all included — word processing, spreadsheet, presentation, video conferencing, and a whole assortment of other apps. These apps have further enabled the ability of employees to work from anywhere and to perform activities that previously required being in the office, or at least attached to an office network in some way. Someone working in Microsoft 365 from their home in New York, for example, can edit an Excel file, which is saved in SharePoint. Another employee from the same company who works out of his or her home in San Diego can then open that same file and view the updates to the file, and maybe they have a video conference with the other employee via Teams to discuss it. No office needed. Totally virtual. You can achieve equal or better productivity than in an office, and you don't have to pay for rent, travel, etc.

The value there is that you can do more than just talk or see one another. You can share word processing documents. You can share spreadsheets, you can share artwork, you can share videos. For example, in the music business and in the film business, they are able to have people with different business units around the globe collaborating. You can have a musician in New Orleans, Louisiana, playing saxophone, and somebody is recording them in Los Angeles in the studio. You can have a film production unit in Australia shooting and then sharing the video with people in New York. I had a direct experience with this because I created a video recently. From my home office I worked with a videographer in Massachusetts. I just sat in my home office in New Jersey by my smartphone recording my video and sent it over to her and she did all the edits. Neither one of us left our respective locations. It was all done remotely, in real time.

When it comes to the internet, which obviously makes all these mobile technologies possible, the connectivity has become much more robust and much more reliable. Internet access has certainly expanded greatly, although it varies by location as not every location has optimal internet availability. Progress has been made, and more progress will be needed. There certainly has been a great advancement, particularly with the ability of fiber internet and affordable options like FIOS (fiber optic, high-speed internet). If you're fortunate enough to be in areas where it's available, you can get pretty robust internet at a very reasonable cost.

To be clear, when we talk about robust internet, we are referring to the fact that the speed is greater for the transmission, and the signal is more stable. Therefore, you can show video, for example, and get it accurately rendered without distortion. Again, not every internet connection is the same. You have fiber-based internet and you have cable internet. Fiber is always better. With cable internet, it's a shared medium. You are sharing bandwidth with people in your neighborhood so if somebody down the block is downloading a big file, or they are watching a video, then it can affect your internet. Fiber, on the other hand, is dedicated to you, so you get all the benefit of the bandwidth that you are paying for. If given a choice, always choose fiber (i.e., like FIOS) over cable. Unfortunately, not all areas have equal options (i.e., no FIOS access) as this is largely controlled by the carriers in a given area and their service plans may not be all that great.

There are often ways to get better bandwidth (i.e., a dedicated fiber line) into hard-to-access areas, but they could get very expensive. These are things to think about if you are considering a move to a particular area — what are the options for internet access? Do they have affordable fiber options, or do they only have cable? Or do they have anything at a reasonable cost? And if not, is there a way to get them and what would it cost? I know someone who lived in an area where Comcast cable service was available, but when they came to the area and were laying cable for free to people's homes, the previous occupant of the house had refused to have them run the cable to his house. Thus, my friend and his wife had to suffer with DSL service, which was painfully slow. They ended up paying Comcast several thousand dollars to connect their home — a service they would have done for free for the previous occupant. At this stage they very gladly paid for it and were happy they did! It shows how dependent we are now on high-speed internet!

Wireless is another area that we should address, whether it's the WiFi in your home or office or whether it's a hotspot. To begin with, wireless technology allows a much cleaner set-up in your home or office, not having to run wires everywhere, especially in areas where it is extremely difficult to do so. In truth, however, wired is always better than wireless in terms of having a stable connection. But, if done right, wireless can be very complementary to your network — and with lower costs than having to run wires where it's not possible and/or overly expensive to do so.

Typically, you want to have access points, and not rely on your router to manage your wireless. Mounting access points where needed in your home and office allows for greater flexibility. If you are working on your laptop, you can pick it up and just walk around and not have to worry about plugging it in somewhere. If you are on the road, you can pick up a hotspot, interact with work, and then get on with your day. I generally don't recommend using public hotspots, however, as they are prone to all kinds of security issues. A much more secure way is to use what is called a mobile hotspot device. You typically get these through your wireless carrier. They are better than using a public hotspot because you are not sharing the connection with anybody and the connections are generally more secure as well. You can take your wireless with you wherever you go and just carry your mobile hotspot with you. You can even share the mobile hotspot with others of your choosing — it is like having your own mobile router, complete with internet connection.

If you do not have a dedicated mobile hotspot device, you can likely use the wireless connection of your smartphone because it typically will have a mobile hotspot capability. However, it may not be quite as good as a dedicated mobile hotspot device, because it is limited by the capacity of the smartphone, which supports a number of other functions. However, it's something you can use if you have no other options where you are located.

While you are limited by your carrier's footprint, mobile hotspots have provided for greater flexibility, allowing you to be more productive in situations where otherwise you would not have been able to do any work.

"Anything involving internet — any device, any type of infrastructure that is on the internet — can be hacked."

We can't stress this enough. Like all technologies, there are vulnerabilities inherent in mobile technologies. Let's speak to some of the vulnerabilities of VoIP for example. While it may seem obvious, you do have to think about security with VoIP and the security of your VoIP provider. You must keep in mind that anything involving the internet — any device, any type of infrastructure that is on the internet — can be hacked, even VoIP. Your provider will typically have infrastructure in a data center, managing the VoIP connections. That infrastructure can be hacked into. You need to evaluate the security of your VoIP provider, making sure that they are doing everything to minimize the chance of them being compromised. Because if *they* get compromised, the threat actors could turn around and exploit all the phones that are connected to the VoIP system and find a backdoor to get into your office through your phone — a good reason to separate your VoIP network from your production network.

You may wonder about collaboration tools, whether there are any specific vulnerabilities to those. Like any app software, there typically is potential for vulnerabilities. Any type of software, especially if it is a cloud-based application, is something to be concerned about. You want to make sure that the developers for your application have a security mindset when creating the application, not just thinking about the feature set, but developing for security as well. As we stated before, very often what happens with software developers is that they just want to get very quickly to market so they are looking to develop software and build the feature set without a thought about security. Then they try to bolt security on after-the-fact, which is an extremely bad design practice.

Security needs to be done hand-in-hand together with development. That is something that developers need to train their engineers to do — not just think about what features they should be putting into the software, but how are you going to make this more secure as you are developing it, not patching it after the fact. You should evaluate the people who are developing the applications and software tools you are working with. Make sure they are following proper protocols for security, not just thinking about features.

What about potential security concerns in video conferencing? From a software perspective, you need to make sure that you are applying updates to the software on your device on a timely basis. And look at the underlying platform. How is data transmitted? Is it encrypted? What is the level of encryption? Where are the video conference provider's data centers located? What path does the transmission of the video conference data take? Does the data stay in-country?

Once again, it comes back to security at the core. Every application developer needs to have security at the core of their application development process. Yes, feature set is important, but it must be coupled hand-in-hand with security.

It is important in your business to choose the right software applications to fulfill your business needs. It should not be done "ad hoc" as you may not achieve the desired outcome. As we have said before, when choosing an application make sure the application is still in active development and supported by the developer, and most importantly make sure that security is kept in mind in the development process.

Businesses have become more mobile and flexible through a combination of technologies including more robust internet availability, wireless, VoIP, video conferencing, and cloud-based productivity applications, allowing for greater collaboration regardless of geographical location.

What is your business doing today to take advantage of these developing technologies?

Chapter 8
Supercharging your Business with the Best Tools

AI: The Next Frontier

At the time I am writing this, in June 2023, AI (Artificial Intelligence) has become a very hot topic. AI applications like ChatGPT are increasingly popular. It seems as though every problem can be solved with AI. And there are the detractors as well, who suggest AI will take over the world and the like. The reality is that, as a technology, AI is in its infancy and it is nearly impossible to predict where it will go in the future. And, like all technologies, it can be used for good or bad purposes. What we can say is that it has great potential to help businesses be more productive and grow. Let's consider how businesses can take advantage of it and what some of the pluses and minuses are of AI.

One of the advantages is that AI relieves humans from some of the more mundane interactions that they don't necessarily need a thinking human to be involved with, which means the business can handle more volume. If you visit a website trying to find service or do some type of inquiry and a chatbot comes up that wants to interact with you, it's probably AI driven. Chatbots can handle some very basic questions in regard to customer service. They will have some canned responses that generally will work for common inquiries so they can handle most items. However, there are some downsides to that as well. Sometimes your interactions and what you want may fall outside the scope of the ability of a chatbot. This is similar to call routing systems — you cannot always get the answer to the question you want without getting transferred to a service person. It is no different with a chatbot. I once had an issue with delivery of what I ordered on Amazon. The first option presented was chat, so I engaged the chatbot. It asked questions and I responded to them, but the help that I needed fell outside of the ability of the chatbot. Once it was clear that it would not be capable of servicing my request, the chatbot automatically transferred my request to a customer service representative, who was able to take care of my request. The takeaway here is that AI is great for servicing a large volume of common requests, relieving the human from the more mundane tasks, and is programmed to escalate those items falling outside of its scope. So, it reduces the potential number of human customer service representatives that you need, but does not eliminate them entirely. The human customer service representatives only work on the higher level, more challenging service issues.

"You didn't take the AI output at face value. You still reviewed it and made sure it made sense."

Another area of AI that you commonly hear about is that it can do your writing for you. No, I am not using AI to write this book, though I may use it to transcribe what I have verbally communicated. So, the content created is my content, but the AI frees me of the burden of having to type out every single word. In that respect, AI can help to facilitate your writing. You still are the original creator, but you simply leverage AI to make your task easier. With that said, it is very common for business professionals to use AI to largely write all of their work. This can be beneficial if done the right way.

For example, you ask your AI system to write an email to one of your customers. You tell the AI that you want to let the customer know of a price increase and outline all the value that you provided to them and enhancements to the service that you are providing. The AI spits out something that is very professionally written, you review it and make minor edits, and send it out. Multiply this by the number of people you send emails to in a day. You spend less time having to write emails, freeing you up for more productive tasks. You didn't take the AI output at face value. You still reviewed it and made sure it made sense and made minor edits to it. Multiply that by the number of business professionals leveraging AI in this way. Think of the enormous productivity gains as a result.

Marketing is another area where you see additional sets of AI driven applications. It is very common now for businesses to use AI to write social media posts or even entire blogs. Certainly that is a great time saver, and they can increase the number of social media posts and blogs, which in theory should boost the level of visibility for their businesses. The potential here is enormous, but there is a caution here as well. You need to be careful of the content that is generated by these systems. As with the email example, you cannot take what is generated and simply copy and paste it onto a post verbatim. You need to review it, make sure it makes sense, make sure it is accurate and factual, and most critically make sure it is not plagiarizing someone else. Keep in mind that AI bases its output on how it was trained and what information it has compiled from the content generated by people. It is not outside the realm of possibility that an AI system could simply spit output for a social post or blog taking material word-for-word from someone's published and copyrighted content. You need to review the content and make sure at the very least that you are not stepping on someone else's toes through plagiarism. You can certainly review these yourself, but there are software apps available that can scan through the text to identify any plagiarized content and rewrite it as needed. You may want to do some combination of both, say, running the scan/auto-edit, then you do a review afterwards and make any appropriate edits. Use AI to review and correct AI, but still have the human review at the end.

AI also has great usefulness in cybersecurity, both for the information security professionals and for the threat actors. Again, as with all technologies, AI can be used for good and bad intentions. Threat actors are now using AI to generate malware on the fly that can potentially evade common endpoint detection and response tools. This lowers the bar greatly for those who aspire to get in on the ransomware boom and have little or no hacking experience. This spells trouble for businesses and other organizations (as if they didn't have enough to worry about).

The job of keeping the threat actors out is becoming ever more challenging due to AI. With that said, the information professionals can also greatly leverage AI to increase their ability to identify and respond to cyber threats. In the course of monitoring a computer network for security threats, a large number of logs are generated. These logs are extremely critical to the function of an information security professional. In these logs there may be subtle hints of suspicious activity. The challenge involves an information security professional sifting through and reviewing every log. It would become enormously time consuming, not to mention very expensive.

AI to the rescue! AI helps information security professionals by doing the grunt work of sifting through every log, identifying the ones that are simply typical network "noise," and those that may be more suspicious. And once the

AI identifies a log indicating suspicious activity, it analyzes it, adds background information, and presents it to the information security professionals in a form they can easily review and decide as to whether it is something that merits further investigation or not. With the current shortage of information security professionals at the time of this writing, AI is a game changer, allowing cybersecurity teams to do more with fewer people and improve their ability to respond to cyber threats. Like all technologies, it is an arms race between good and bad — in this case between the threat actors and the information security professionals.

The potential benefits of AI to businesses are endless. We are early in the game and as the technology continues to develop, it will get better and better and likely require less and less human intervention. I would throw out a caution, however. The common theme here is that AI can do the vast majority of a given task, perhaps even 90-95% or more of the work potentially, but you still need a human to review the output. Do not simply accept what is being produced or allow the AI to make the important decision for you if it is something that requires a review of output and a decision. It requires a balance. As with all technologies, the purpose is to take the human out of the mundane tasks and allow us to focus on the more interesting and high value tasks. Business owners struggling with what to make of AI need to consider how to make the best use of the benefits of AI to drive efficiency and growth while making sure they are managing the process and keeping humans at the tail end of that process for final review and direction.

Are you using AI in your business?
If not, have you identified areas in your
business that may benefit from AI?

Chapter 9
Getting Started on a More Strategic View of IT

How do you get started on the strategic model of IT? It all starts with you and how you view things as it relates to products and services that have an impact on your business. In a business, it is the executive management team that sets the tone for the organization. It is the difference between viewing IT as a strategic asset to your business versus the IT "janitor" that fixes problems and tends to the infrastructure as needed, much like the person that waters your flowers. The person watering your flowers, however, typically does not have a direct impact on your business operations. If they forget to water the flowers in the office one day, the flowers may wither but your business will continue (unless of course if you sell flowers!). On the other hand, if the IT person doesn't perform planned maintenance on the server, and the server crashes as a result, your business WILL be impacted, perhaps even severely. Whether you have an internal IT department or an MSP, you need to view it as a partner with your business that will collaborate with you to deliver definitive business results. Without buy-in from top management, any strategic initiative will surely fail.

The next step is to think about exactly what you want to accomplish in your business, because everything you do is driven by your business and its operations. What are the critical data flows in your business? What are all the different steps in that process? This gives you an idea about who you are going to need to have on board with you.

Once you figure out what your business goals are, you will need to determine what expertise you are going to need to either build or modify your IT infrastructure/systems to accommodate that. You need to get the right team in place. How do you do that? And what constitutes the right team? How we define teams typically depends on the size of your company. If you are a larger company, you may hire all your IT people as employees. Conversely, if you are operating a smaller business with limited resources that team may be outsourced. Or it could be a combination of employees and outside providers as part of your team.

Let's revisit the subject of insourcing versus outsourcing. Other than the fact that a business might be small, how do you know which way is better for you? It's going to come down to scaling. Consider that hiring an IT team brings associated costs with paying salaries — from staff members to the manager to the IT Director, and, if you include the executive level, the CIO. All of those costs plus some of the extended services that you may need like cybersecurity can add up pretty quickly for a smaller business. Ultimately, it's going to come down to what your budget is. It's certainly preferable to have everything in-house because you will have greater control and you will have the loyalty of all the people you hire. If your desire is to have control because you are uncomfortable with having IT contractors, outside parties that have their hands in your IT, then you need to hire internally provided you have the financial resources to do that.

You may also have security concerns. If so, you want to have as few outside vendors as possible. Then practically all of your IT staff will be hired employees. There may be some businesses that are in highly sensitive industries — government contractors, for example, that may have to really limit access to their data. If you are in a very sensitive business, that is one of the catch 22s. You have to be very careful about bringing on outside vendors because they have to have all the proper security protocols in place. Anybody that touches your network must have all their ducks in a row in terms of the required cybersecurity standards. Your own security has to meet the standards set by the current cybersecurity standards that you need to adhere to, and by extension your vendors need to adhere to them as well. Evaluating every vendor's compliance to cybersecurity standards can be very time consuming and challenging to do.

"Most importantly, they must closely match your company's core values."

In regards to hiring the right IT team internally, how do we determine if somebody is right for your business? How do we determine whether they will support that business mission you are looking to achieve? To begin with, you might want to see a track record of success stories and a history of companies they have helped, what they have done there, and what the results were. You might look to a larger extent at their technical expertise, making sure they have the technical knowledge for their roles. However, non-technical/soft skills are important. They need to be of a collaborative mindset, willing to work as a team with all of the other IT department members, not to mention with the rest of the management and staff. At the end of the day, you need to see what they have done for other companies, how they helped them, what the results were, and, ideally, if you can get references, or feedback, or other helpful information from outside parties. Validate and verify their stated achievements. And remember, most importantly, they must be of a collaborative mindset and closely match in terms of your company's core values. Without these, all the technical background does not matter and they may in fact be counterproductive to your business.

Sometimes it is simply not possible for you to bring everybody in-house. We have mentioned in this book that even if you have an IT staff in-house, there may be times when you still must outsource certain functions. That may happen if your IT staff is overworked; if they are spending 99% of their time putting out fires instead of working on projects that can potentially improve your network, they do not have time to do anything strategic. There are also times when your company may require specialized expertise that you just do not have in-house. If you are working on a special project, your staff in their day-to-day work may not have expertise in a specific area so special projects may be an area where you want to bring in outside IT consultants. It might be a one-off just to do that project, or, eventually, for multiple projects, or it can be more collaborative, and ongoing, where the outside provider acts as almost an arm of the IT department, working hand-in-hand with your internal IT staff. In that case both work together almost as one department where the in-house staff focuses on the day-to-day troubleshooting while the outside IT provider may focus more on the larger projects and augmentation for any issues that may be beyond the capabilities of your in-house department. It becomes more of a symbiotic type of relationship. It is a very common misconception that an outside IT provider will have an adversarial relationship with an internal IT department. Technically that could result in the internal IT staff being replaced by the outsourced provider. While that could happen, in an ideal scenario they work in tandem, and the outside IT provider in fact helps to ease the burden and allows internal IT people to do their jobs better, which is a win-win situation for everybody.

You probably would need somebody to act as a CIO-type person. That could be an employee, or often in a smaller business that role will be outsourced as a separate consultant. You must have someone who is the point person for all things IT related, and you want the focus to remain strategic, not technical, because the mistake that is often made is to just go out and hire an IT manager and maybe a few IT staff members to build your infrastructure, maintain it, and troubleshoot. That's not the best way to do it because the folks will be technical and not strategic. They will just go and buy resources for the sake of making sure things run properly and minimize computer problems. They are not thinking about how it's going to support your business processes and objectives. Look for somebody at a level who is strategically oriented, and then underneath him or her have an IT director — and depending on the size of your company, maybe an IT manager. Once you have the higher-level managers, use them to help hire the rest of your support staff. You, as a business owner or manager, will delegate responsibility to someone who will supervise IT staff, coordinate end-use support, and work with them to execute on IT initiatives that will bring the business forward. With a strategic focus at the top, the team will support the strategic direction of your business.

Is hiring a complete IT staff outside your budget? You are not alone. That is a common theme with small businesses. You do not need to hire all of your IT team internally. Outsource! MSPs are the answer! How would you go about hiring one? What would you look for if you were interviewing and you wanted to hire an MSP? How would you decide whether they are the right one for you? First, ask how many years they have been in business. What commonly happens is that you have IT consultants that are between jobs. They are just using the consultancy as a transitional period to earn some money until they get a regular job somewhere else. That's not the person you want; you must be careful to make sure the person you are hiring is not that kind of person because you want a long-term relationship with someone who is fully focused on his or her business. This needs to be a strategic, long-term relationship so make sure it's a company that has been in business for at least five years as a rule of thumb.

"Make sure they have enough bandwidth to service your needs."

Number two, while there are many IT consultants who are solo practitioners, make sure that you are not dependent on one person. Now, I'm not saying don't hire a solo practitioner, but make sure they have other people who work with them — and at the very least they partner with other IT professionals. Make sure they have at least three or four people on their team who are available for any troubleshooting, so that you are not dependent upon the owner of the company to come in and do all the work. You want to make sure they have enough bandwidth to service your needs.

If you are a business owner, you are not going to just evaluate technical experience. I would check the references of people who have used them and it's even better if they work in your industry — but certainly get some kind of testimonial or other feedback from businesses that have used that particular IT company and verify the work they claim to have done. When you are talking with them, find out what their processes are and notice if they have an interest in your business — if they are asking you questions about your business and seek to understand it — rather than just giving you tech speak. You want an IT provider who understands your business and is not just selling a piece of tech equipment or merely looking to troubleshoot your tech problems — you want a provider who understands your business and your business needs and will look to service those needs wherever appropriate.

Part of the journey of making your company more strategic in its use of IT involves looking at your own current business processes and adjusting them. Let's take a look at how you would go about doing that. You have to first take a 10,000-foot view of your business and do a really deep dive. What is your business purpose? How do you make money? What are the business processes driving that? How does data flow inside your business? Not just conceptually, but how it flows through the different processes that are in your business. What technology infrastructure is supporting those processes, whether on-premises or in the cloud? You might find things are inefficient in some areas. This is where you can identify how data is moving in your organization. Is it moving in the most optimal way, in the most secure way?

What's the technology underneath that is supporting that? Let's take an example of what might be a situation where somebody evaluates what is going on in their business, and they find holes, so to speak, where they find blockages that need to be addressed.

Suppose we consider a CPA firm. They use a whole myriad of different applications to support their various processes. Let's say, for whatever reason, that when they are processing tax returns, they use two different types of tax software and there is one group of employees that use one software package and a second group of employees that uses the other one. The two groups don't talk to each other and can't share information with each other, so the collaboration is limited between the two groups. That's extremely inefficient. They are paying money for licenses to each software package and Group A can't speak to Group B or vice versa and can't share or collaborate on those returns. So, the solution is: let's just take a step back, define our procedures for processing tax returns and pick the best one of the two software packages to support that or maybe get rid of both and get a package they all use and migrate to the common package. Now they can all work together, they can share tax returns, they can get work done quicker and more efficiently and service clients much better. They unified not only the software package, but the process that it supported. The process really drives everything. We unify the process, and a swath of underlying technology (the software) using one tool set instead of two. so now the process runs more efficiently.

In some cases, the impediment may be either a lack of training or resistance to using something new. It could get very political; everybody has their pet preferences. If one set of partners prefer to use only their chosen software and the other group of partners feel the same about theirs, then the initiative doesn't move forward because the people that run the firm can't agree to unify their processes to make things more efficient. Politics gets in the way of technical efficiency and the underlying processes that support them. That's why all the people that have a decision-making role in the company need to be on board for making things more efficient and put aside any pet preferences for one software or solution over another.

Let's look at a manufacturing business and explore how you might examine the processes and adjust them there.

What might be some of the areas where they are experiencing difficulties that need to be adjusted in their processes? It is a matter of knowing, again, the flow of information, what information people that are on the floor need to do their jobs. What information needs to be fitted into the operational technology that will support the production — even from the standpoint of ordering raw materials, being able to track the workflow from the time the materials are ordered to the time the materials arrive, to the time the job is started to the time the job is completed. There are so many moving parts here. There could easily be a problem with, for example, silos in the organization, a lack of communication amongst them — such as inventory and production.

It always comes down to the flow of information in your process. If you have silos of information, you don't have a unified system for handling the information flow in your organization. You can have employees doing one thing in one part of the company and others doing another thing — another part that's not talking to you. Nobody knows what to watch for. When you need to reorder something, the people that take the orders may not be talking to people who are in the warehouse with the materials. That's why you have ERP systems (refer to Chapter 2) that unify all aspects of the company, including the production area, as well as the operational and the financial areas of the company. Having one unified system to track everything at all levels will accomplish that. So, by unifying the process, the line-of-business software will help facilitate the process for you to make informed business decisions. Getting rid of all the silos will unify everybody on well-defined process flows that will be supported by the ERP system (the technology), which facilitates the flow of data throughout the organization, including executive level, production, inventory, manufacturing, sales, and wherever else it may be needed.

**"Create a map of all your functions
and then how the information is
flowing through those functions."**

Let's suppose your business is not quite that sophisticated, and you don't have an ERP system. You want to look at your processes; you want to adjust them, so what do you do? Do you do this on a whiteboard? How would you trace out what your processes are? Take a step back and look at all the workflows in your company. In this case, in a manufacturing environment, how you make money, basically all the steps you go through to make money — getting the order from the customer, producing the widget, or whatever combination of parts that you need to assemble that one product. Then multiply that by however many orders you are getting in and easily communicate information to the people who are doing the production, the packaging, warehousing, and shipping to the end customer. You need to create a map of all your functions and then how the information is flowing through those functions. You do a workflow diagram of all the process steps in your company, and all the data flows that are supporting those processes. You can do it on a whiteboard; you can write it on a piece of paper — just get it in some documented form. Then you can hash it out and share it within your organization. This should be a collaborative endeavor, with all the key areas of the organization from the CEO all the way down to all the managers and staff in the various departments. Once you have it all together, you want to look at areas for process improvement — even before the technology aspect of it, look at the process.

Let's take a look at how that works. Once you have documented your workflows as they currently exist, and you have determined what your desired workflow looks like, then you can start looking at what solutions may potentially work best to support that. To an extent, it may depend on what type of industry you are in, but at a base level you talk to the vendors about what you want to accomplish. Choose the ones who understand the goals of your organization, what you want to accomplish, and will build a system around that will best serve the business needs. Not the other way around. Never go to your vendors first, because they will be biased into wanting to just sell you their system. You determine your processes first, modify them as desired; then you have a firm understanding of what you need, and you fit the technologies to your needs

You may be wondering when you should reevaluate the systems and determine whether they need replacing or upgrading. Once again, to an extent, it does depend on your industry and the type of technology you are talking about. Irrespective of those factors, it's going to be a continuous process of evaluation — the job is never done. That process may move a bit faster or slower depending on the size or type of business you are in. There are going to be continuous evaluations over time, taking into account any changes in the business, with recommendations made for change, which will be evaluated and either accepted or rejected by management. It is never an optimal scenario; it never ends, especially as the business itself changes, which in turn may affect your processes, and in turn your technology needs.

As a business owner, you need to differentiate and determine what is a technology want versus a technology *need*. You must always take a step back and ask why are you buying this; what is the improvement that this purchase solves? What is the benefit to the business? What problem is it solving to make the business better? If you are buying a new server, a piece of highly technical and relatively expensive equipment, what problem is that solving for the business? Perhaps the server it is replacing is a very old server, and maybe subject to frequent maintenance; it has slowed down and it is hindering the flow of production because of that, not to mention the operating system is no longer supported by the developer, opening you up to a security exposure which would be potentially highly disruptive and costly to the business. In this case, you are solving multiple problems by buying this new server. By buying a brand-new server, it's going to have greater uptime, it's going to run faster, people are going to get the work done faster, production is going to move faster, and it will be more secure. Also, it may have better storage capacity to support future growth. Ultimately, it's going to have some effect on the bottom line.

"Plan your purchases to support the company's objectives; tie them into a business result."

Every product has to be evaluated in this same way to see what business result will come from purchasing the item. Evaluate every purchase — if you are buying a desktop computer, why are you buying it? What is the reason — buying a desktop that is going to work much quicker, getting work done faster, and so forth. Why are you buying 10 computers? You are going to be hiring 10 company employees and those support the growth of the business by supporting those employees. You need to plan your purchases to support the company's objectives, tie them into a business result. Avoid buying this computer because you just want a newer machine. Don't just buy a MacBook because you like Macs when your whole company is PC based. That may be the wrong application because now you are trying to fit a square peg into a round hole. There may be incompatibilities with the application, and getting two different platforms to talk to one another, etc. You may end up making things less efficient and more costly as a result. Avoid the shiny new cool object syndrome!

You may be wondering, with technology changing as fast as it does, how does a business stay ahead of the curve as part of its strategic application of technology? First, consider that your time is limited. Certainly, you can find out a lot from the internet, but you have to be careful about that because of the latest and greatest technology out there. There are shiny new technologies every day, and how do you know whether they are the best application for your business — or maybe something that really isn't ready for production? That's where you need to rely on experts. Talk to either your IT department to get feedback — or your trusted IT consultant/provider. That is their job. You can look at reviews in the trade press, which may prove helpful as well — with a caveat not to fall in love with the new shiny objects as I mentioned earlier. If you see the newest thing in your industry, and it's really getting rave reviews, be careful because you may not necessarily want to be the early adopter. If you jump in too early, you may very well have a very bad result — the opposite result of what you intended by buying that shiny new object you saw in the trade journal. Let it prove itself in the marketplace. Then maybe after six months to a year, consider adopting that technology in your company, but only after careful assessment, talking to others in your industry that have used it, etc. As we warned you earlier, it bears repeating — be wary of overly aggressive and persuasive salespeople from startup companies who tell you about all the wonderful features of their product and that you should get in on it now. Stand firm in following the process of business needs evaluation and using proven technology — your business will benefit in the long run.

> *Are you ready to get on the road to a strategic approach to technology?*

Epilogue
What Might Near Future Technology Hold?

With an eye to the near future, based on recent trends, here are some thoughts about what may happen that could affect your business, and suggestions on how to react.

Let's start with AI, which will probably become a good part of the future. While it's controversial in some respects, it will probably be more influential than we thought it would be. At the same time, there's no question that it's probably going to get reined in by regulators at some point. It's like any new technology. There are beneficial use cases and companies are springing up right and left to bring forth AI related technologies to the marketplace. It's something that all businesses should be looking at to help their businesses to be more efficient and grow--with the caveat that you don't let it take over your process. The people in your business still need to control the technology, not the other way around. As I mentioned earlier, make sure there's a human factor at the last mile of every AI technology you put in place.

In addition to the AI, how about the Metaverse? What possible ways would the Metaverse apply to small businesses? I think that there are a lot of possibilities with the Metaverse. Bigger players have tried it — most notably, Facebook, which spent $36 billion on the Metaverse. However, the big players are pulling back. They have found that the technology was awkward. It just somehow has not worked out. I don't necessarily think it's something that's going to fall by the wayside. It's probably something that's 10 or 20 years out before the technology catches up and becomes a bit more refined and palatable to the average person. That may make something like the Metaverse a more viable platform. Is there any foreseeable potential use for the Metaverse by small businesses? In the near term, probably not. Non-tech users typically have found it very awkward and cumbersome to work with. What we have seen people currently using is still in the alpha stage, a very preliminary, very buggy stage. When it gets there, it potentially could have applications for businesses. For example, you could have a virtual meeting within the Metaverse as if you were all together in the same conference room, instead of having to do it through Zoom. But it may become an evolution of Zoom or similar applications to that type of platform. If it works in its ideal form, it will put you in a whole other world. It will be very seamless, like you're on another planet. You'll be able to interact with other people as if you were there. You may want to go into virtual stores or virtual offices, walk virtual sidewalks, drive in virtual streets etc. You might call it an example of augmented reality. It's kind of a way to manage reality. That's really the ultimate. The Metaverse, as its name may implies, is its own reality.

We have seen some interesting examples of how people are using mixed reality in business. They're using a combination of reality and augmented reality together, a hybrid. You could have a virtual meeting, or you could be sitting in a conference room and on the other side of it is a virtual world. We're all interacting with each other. One in the real world, one in the virtual reality of the Metaverse-type world, interacting with the outside world. For example, a company called Revealio has an application where you could take a book or some type of marketing material, and if you scan it with your smartphone, you see something come to life. I could see a lot of different applications with that. I am thinking, if somebody shows a hologram of somebody else interacting with you as opposed to having a conversation over the phone or via Zoom. Your holographic version of the person that is sent is in the room with you, and they can see you in a holographic representation on their end. That could have applications for product demonstrations and new product introductions. You can have a virtual copy of your products demonstrated in another location without having to drag your hardware around if it is large and cumbersome. You would see all three dimensions of it being demonstrated over in the other location. While that's pure speculation on my part, that would be amazing. That technology does exist, it is just a matter of whether it is affordable, technically mature, and accessible enough for businesses to use.

For example, look at how the exercise companies are using technology to have an interactive experience for the person at home doing exercise. There is a trainer there, watching in real time on screen, giving the user guidance. We have seen the rise of telemedicine, which of course, became very useful during COVID-19 when people were home confined, and they could still interact with their healthcare providers. Now we have all these smart devices for health monitoring that people can wear and interact with their health care provider remotely. Imagine how all of these could evolve. Your remote personal trainer could almost be literally in the room with you coaching you through your workouts. A virtual image of your doctor could be interacting with you reviewing the scans made on your body by the smart devices. Perhaps you may not even need blood work anymore! Imagine knowing your key body statistics in real time without having to wait for a doctor's appointment, and being able to take corrective action immediately.

There is potential application in other fields besides medicine, like education, for example. I am envisioning worldwide courses from a university where you have immediate feedback, interaction and test scores. You can have virtual classes, and again, it goes back to virtual reality. The Metaverse could be leveraged to have classes with the look and feel of a live class. The instructor would be in one location while students are in classrooms across the world, perhaps even in their homes, taking a class and interacting with the instructor and other students. The same thing could be applied to business education. You have a busy person running a business that does not have the time to be going to classes on a physical campus, but they can do so virtually. In the IT industry, for example, you have conferences all the time where people are going to see speakers, interact with vendors and colleagues, and see demonstrations of the new technologies to keep on top of what's out there. Just imagine if that is all virtual. A lot of time goes into having to travel, get to the hotel, get a room with all the inconvenience of that, not to mention the cost.

We will probably never completely replace the conference/trade show environment. We are still human, and we like to have personal interactions. You saw that after the whole COVID-19 period. During that time, everything went virtual; there were virtual conferences, but the technology was not refined yet. It was a bit cumbersome. As soon as things reopened, in-person conferences popped up right and left. People like getting together and interacting in a social setting. You may very well see hybrid-type conferences, where some people attend in person, while others attend virtually, with the ability to "walk around," seeing holographic images around them, while people at the conference can see them in similar fashion. Wouldn't that be cool?

"We are making it more experiential — as opposed to transactional."

The question remains, how will people continue selling products and services in the future? As it is, because of the internet, things have become very transactional. In many industries, we have taken the human element out of it and things are done strictly on a cost basis online. With that said, the basic foundations of sales do not change. The platforms that you use change; they may evolve over time. But the fundamentals of sales do not. Maybe the newer technologies will enhance that aspect. If you are selling clothing, having a picture of the clothing with an actual holographic representation of it that you could interact with, for example. There is already technology where you can see what you would look like wearing a certain piece of clothing. The same way with home décor; you would see what your home would look like if you chose a certain wallpaper. They can adapt the color paint you select to your picture of your room. It makes the buying experience much more interactive and alleviates some of the resistance to buy online without seeing the results. The reality is that the direction we are going is not to take the human out of the selling experience with technology, but rather to enhance the experience leading to a transaction. It's a basic principle of sales; anything that takes away resistance, or potential objections to the sale, will result in more sales. In short, we are making it more experiential — as opposed to what's traditionally been e-commerce, which is just transactional.

We can expect to see a continued evolution of cloud-based technologies serving small businesses and small offices. We probably will not see servers anymore in small offices with fewer than 10 or 20 employees. If they have not gotten rid of their server equipment by now, they most likely will soon. It's more beneficial for them to just go all cloud and do away with the expense and complications of running a physical server in-house. You may see this to a degree in larger offices as well, with whatever servers they do have running virtually in cloud environments. This is a bit harder with larger server infrastructure, but it is achievable over time. Cloud technologies still need to evolve to make it easier, better, more cost effective. Applications will also likely become more cloud enabled, evolving to Software as a Service (SaaS) applications, in which case you would not need to run them on your computers or infrastructure. You are seeing that now, and I predict you are going to see more of that in the future. Computers will become more like dumb terminals, just accessing all the applications over the internet. Ironically, it would be something of a throwback to the old days of mainframes. There will always be those that do not trust the Cloud, and will want to only run applications stored on their computers, but it may become harder and harder to do that.

This again leads towards what we have been saying about the diminishing need for office space, which then leads back to more remote work, which then begs the question about home offices--which creates a whole new business opportunity. Building enhanced home offices. That may be the future of my industry, the IT industry, supporting home-based offices. The typical MSP has shied away from residential clients, which have not been attractive or cost-efficient model. But that might become more necessary. Consider that in this scenario you are looking at managing multiple home offices, so it's really not that much different than what they were doing with a large office situation. It just involves more locations. The technology might evolve, again, maybe with some help from AI for more of a self-service type of model for home users to have security in a box, so to speak. Maybe your employer's platform will provide the necessary tools.

Taking a cue from the smart home automation industry, perhaps that same technology could be adapted to an MSP managing multiple remote offices for one company. That is pure speculation. In truth, it is a problem that needs to be solved. You have large numbers of remote home-based users who are reluctant to come back to the office. How do you solve security, and make sure that they have what they need to work efficiently without potentially compromising themselves and their employer? An important aspect of that is separating their personal devices from their business environment. So the kids aren't using the same computer for their gaming that you're using for work. You may have a plug and play router, which automatically creates a home wireless network and a home business network, and maybe a guest network as well. We have been going in that direction through our routers that will automatically create a home and a guest network. Why not add a third one for a business network? It is a virtual LAN. A LAN basically is a network where computers/devices can talk to each other. You get a company issued laptop and you will connect that to your business network. Half the battle is separation between home and business networks. The more you can facilitate that, the more security you are going to have.

> *How do you think technology will evolve, and how do you think that will affect your business?*

ACKNOWLEDGMENTS

This book would not have been possible without the help and support of a number of great people:

- Marvin Fischer, Don Waters, and John Hauryluke for stepping up to the plate and sharing their stories
- Barry Cohen, my editor for spending a great deal of time pulling the ideas out of me into written form, and his constant "nagging" to keep me on track to get the book done
- Heather Felty for taking the time to go through my manuscript and help me polish it up
- The many people in my network for their support and encouragement

About David Quick
Founder and President
of Total Cover IT

David Quick has been in the IT profession for over 25 years. He always had a love for technology and what it can do for businesses, even before he worked in IT.

While working as an accountant for a local community bank, he took a very manual accounting process, which included pencil and 14 column paper among other paper-intensive processes, and moved them all to spreadsheets, greatly improving the efficiency of his role. After leaving that job, David held several IT/technology roles, with his last full-time position as IT Director of Rosenberg Rich Baker Berman & Company (RRBB), a regional accounting firm based in Somerset, New Jersey. During the 17 years that David was with RRBB, David oversaw the implementation of systems to support the growth of the firm from about 55 people in 2000 to more than 80 people by the time that he left in 2018.

David and his team transformed the firm's IT infrastructure from antiquated, unreliable hardware to enterprise-grade architecture. Downtime was minimized, remote work was enabled, long before it was mainstream, resulting in greater efficiency and productivity, providing a solid foundation for the firm's growth. When the firm decided to move its headquarters location, David led a project to transition the firm's computer systems and network infrastructure to the new office with minimal downtime.

During his time with RRBB, David gained in-depth knowledge of the special needs of CPA firms, with a deep understanding of the types of software applications used as well as the workflows and what keeps them up at night. As founder and President of Total Cover IT®, David now brings his combined business-oriented IT management expertise and CPA firm experience to accounting firms and other businesses that wish to grow and need an IT resource that fully understands their business and takes a business-first, strategic approach.

David's guiding philosophy is "getting the technology right the first time" by a thorough assessment of business needs, detailed project planning and execution, and best-of-breed solutions. This process ensures successful project completion, minimized downtime, and maximum ROI.

David has an M.B.A. in Finance from Seton Hall University and a bachelor's degree in Accounting from Kean University.

When not working on improving the use of technology at his clients, David enjoys fitness activities, including long walks, weight training, Yoga and Pilates, as well as going out for meals with his friends (especially when it's something sweet!).

If you would like to learn more about David and Total Cover IT, please visit the company website at https://www.totalcoverit.com

www.ingramcontent.com/pod-product-compliance
Lightning Source LLC
Chambersburg PA
CBHW071643210326
41597CB00017B/2100